Common CORE Mathematics

Practice at 3 Levels ●●●

Table of Contents

Using This Book

What Is the Common Core?

The Common Core State Standards are an initiative by the states to set shared, consistent, and clear expectations of what students are expected to learn, so teachers and parents know what they need to do to help them. The standards are designed to be rigorous and pertinent to the real world. They reflect the knowledge and skills that our young people need for success in college and careers.

What Are the Intended Outcomes of Common Core?

The goal of the Common Core Standards is to facilitate the following competencies.

Students will:
- demonstrate independence;
- build strong content knowledge;
- respond to the varying demands of audience, task, purpose, and discipline;
- comprehend as well as critique;
- value evidence;
- use technology and digital media strategically and capably;
- come to understand other perspectives and cultures.

What Does This Mean for You?

If your state has joined the Common Core State Standards Initiative, then as a teacher you are required to incorporate these standards into your lesson plans. Your students may need targeted practice in order to meet grade-level standards and expectations and thereby be promoted to the next grade. This book is appropriate for on-grade-level students as well as intervention, ELs, struggling readers, and special needs. To see if your state has joined the initiative, visit the Common Core States Standards Initiative website to view the most recent adoption map: http://www.corestandards.org/in-the-states.

What Does the Common Core Say Specifically About Math?

For math, the Common Core sets the following key expectations.

- Make sense of problems and persevere in solving them.
- Reason abstractly and quantitatively.
- Construct viable arguments and critique the reasoning of others.
- Model with mathematics.
- Use appropriate tools strategically.
- Attend to precision.
- Look for and make use of structure.
- Look for and express regularity in repeated reasoning.

Common Core Mathematics Grade 2 • ©2012 Newmark Learning, LLC

How Does Common Core Mathematics Help My Students?

- **Mini-lesson for each unit** introduces
  Common Core math skills and concepts.

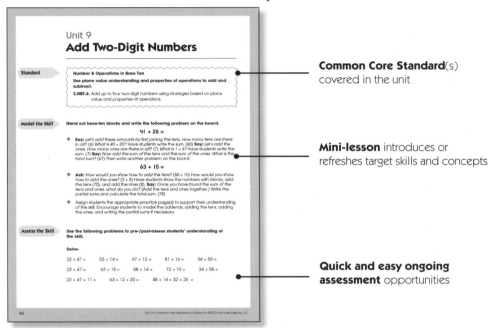

Common Core Standard(s)
covered in the unit

Mini-lesson introduces or
refreshes target skills and concepts

**Quick and easy ongoing
assessment** opportunities

- **Four practice pages** with three levels of differentiated practice,
  and word problems follow each mini-lesson.

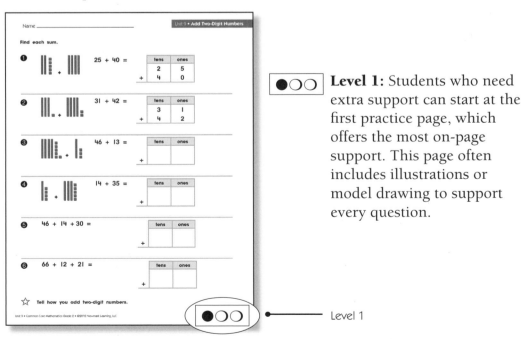

Level 1: Students who need extra support can start at the first practice page, which offers the most on-page support. This page often includes illustrations or model drawing to support every question.

Level 1

Level 2: The second level of practice offers streamlined support features for the first few problems (illustrations, model drawing, or an algorithm reminder for support).

☆ Each practice page includes a bonus thinking-skills question so students can answer "How do you know?" to address Common Core Standards of Mathematical Practice and demonstrate their reasoning and understanding of the concept.

Level 3: The third practice page does not offer on-page support and depicts how students are expected to be able to perform at this grade level, whether in class or in testing.

☆ **Tell how you found the sum.**

Bonus Thinking Skills question on each practice page

Level 2

Level 3

Word Problems: Each unit ends with a page of short answer and multiple-choice word problems so students are challenged to marry their computation skills with their quantitative-reasoning and problem-solving skills and grow more familiar with the types of problems they will encounter on standardized tests.

Word Problem Page

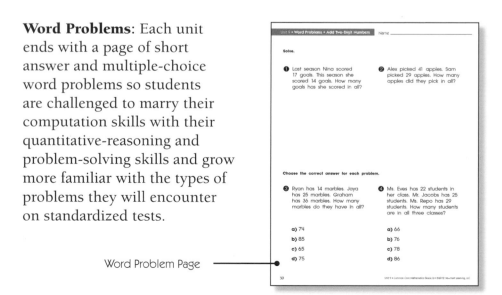

Common Core Standards Alignment Chart

Units	2.OA.1	2.OA.2	2.OA.3	2.OA.4	2.NBT.1	2.NBT.2	2.NBT.3	2.NBT.4	2.NBT.5	2.NBT.6	2.NBT.7	2.NBT.8	2.NBT.9	2.MD.1	2.MD.2	2.MD.3	2.MD.4	2.MD.5	2.MD.6	2.MD.7	2.MD.8	2.MD.9	2.MD.10	2.G.1	2.G.2	2.G.3
Operations & Algebraic Thinking																										
Unit 1: Add and Subtract Fact Families		✔																								
Unit 2: Write a Number Sentence	✔																									
Unit 3: Odd and Even			✔																							
Unit 4: Add Equal Groups				✔																						
Number & Operations in Base 10																										
Unit 5: Understand Place Value					✔																					
Unit 6: Count, Read, Write to 1,000						✔	✔																			
Unit 7: Compare Numbers								✔																		
Unit 8: Use Strategies to Add									✔																	
Unit 9: Add Two-Digit Numbers										✔																
Unit 10: One Hundred More or Less												✔														
Unit 11: Add Three-Digit Numbers											✔															
Unit 12: Use Strategies to Subtract													✔													
Unit 13: Subtract Two-Digit Numbers									✔		✔															
Unit 14: Subtract Three-Digit Numbers											✔															
Measurement & Data																										
Unit 15: Inch, Foot, Yard														✔	✔	✔	✔									
Unit 16: Centimeter, Meter														✔	✔	✔	✔									
Unit 17: Add and Subtract Lengths																		✔	✔							
Unit 18: Tell Time (Nearest Five Min.)																				✔						
Unit 19: How Much Money?																					✔					
Unit 20: Make a Line Plot																						✔				
Unit 21: Make a Graph																							✔			
Geometry																										
Unit 22: Identify Shapes																								✔		
Unit 23: Parts of Shapes																									✔	✔

Unit 1
Addition and Subtraction Fact Families

Standard

> **Operations & Algebraic Thinking**
>
> **Add and subtract within 20.**
>
> **2.OA.2.** Fluently add and subtract within 20 using mental strategies. By end of Grade 2, know from memory all sums of two one-digit numbers.

Model the Skill

Hand out connecting cubes and model the following problem on the board.

- ◆ **Ask:** *How many cubes do you see that are dark cubes?* (4) *How many lighter cubes are there?* (2) Have students make a cube train with 4 cubes of one color and 2 cubes of another color. **Say:** *The cubes show the addition fact 4 + 2 = 6. Break the cube train into two colors. A related subtraction fact uses the same numbers as the addition fact. What related subtraction fact did you show?* (6 – 2 = 4 or 6 – 4 = 2) Have students record one of the facts.

- ◆ **Say:** *Make a cube train for 5 + 3 = ?. What is the total number of cubes?* (8) *Where do you write the total in a subtraction fact?* (Possible answer: It is written first.) Observe as students break their cube train into two colors and record one of the related subtraction facts. (8– 3 = 5 or 8 – 5 = 3)

- ◆ Assign students the appropriate practice page(s) to support their understanding of the skill.

Assess the Skill

Use the following problems to pre-/post-assess students' understanding of the skill.

3 + 1 = 4	2 + 6 = 8	6 + 5 = 11
____ + ____ = ____	____ + ____ = ____	____ + ____ = ____
____ – ____ = ____	____ – ____ = ____	____ – ____ = ____
____ – ____ = ____	____ – ____ = ____	____ – ____ = ____

Write a related subtraction fact.

① 5 + 3 = 8

_____ – _____ = _____

② 7 + 3 = 10

_____ – _____ = _____

③ 8 + 1 = 9

_____ – _____ = _____

④ 6 + 3 = 9

_____ – _____ = _____

Write a related addition fact and subtraction fact.

⑤ 4 + 6 = 10

_____ + _____ = _____

_____ – _____ = _____

⑥ 2 + 7 = 9

_____ + _____ = _____

_____ – _____ = _____

 Look at the page. Draw a circle around the cube train that can show
10 – 6 = 4.

Name _____

Write the related addition and subtraction facts.

❶

8 + 2 = 10
___ + ___ = ___
___ – ___ = ___
___ – ___ = ___

❷

8 + 4 = 12
___ + ___ = ___
___ – ___ = ___
___ – ___ = ___

❸

7 + 8 = 15
___ + ___ = ___
___ – ___ = ___
___ – ___ = ___

❹

5 + 6 = 11
___ + ___ = ___
___ – ___ = ___
___ – ___ = ___

❺

3 + 9 = 12
___ + ___ = ___
___ – ___ = ___
___ – ___ = ___

❻

3 + 7 = 10
___ + ___ = ___
___ – ___ = ___
___ – ___ = ___

❼

5 + 10 = 15
___ + ___ = ___
___ – ___ = ___
___ – ___ = ___

❽

5 + 7 = 12
___ + ___ = ___
___ – ___ = ___
___ – ___ = ___

❾

8 + 6 = 14
___ + ___ = ___
___ – ___ = ___
___ – ___ = ___

☆ **Tell how you find related addition and subtraction facts.**

Name _____

Use the numbers shown to write the facts in the fact family.

❶ 12   5   7

____ + ____ = ____
____ + ____ = ____
____ − ____ = ____
____ − ____ = ____

❷ 15   7   8

____ + ____ = ____
____ + ____ = ____
____ − ____ = ____
____ − ____ = ____

❸ 13   6   7

____ + ____ = ____
____ + ____ = ____
____ − ____ = ____
____ − ____ = ____

❹ 14   5   9

____ + ____ = ____
____ + ____ = ____
____ − ____ = ____
____ − ____ = ____

❺ 13   4   9

____ + ____ = ____
____ + ____ = ____
____ − ____ = ____
____ − ____ = ____

❻ 16   7   9

____ + ____ = ____
____ + ____ = ____
____ − ____ = ____
____ − ____ = ____

❼ 12   3   9

____ + ____ = ____
____ + ____ = ____
____ − ____ = ____
____ − ____ = ____

❽ 11   4   7

____ + ____ = ____
____ + ____ = ____
____ − ____ = ____
____ − ____ = ____

❾ 15   6   9

____ + ____ = ____
____ + ____ = ____
____ − ____ = ____
____ − ____ = ____

❿ 17   8   9

____ + ____ = ____
____ + ____ = ____
____ − ____ = ____
____ − ____ = ____

⓫ 11   5   6

____ + ____ = ____
____ + ____ = ____
____ − ____ = ____
____ − ____ = ____

⓬ 12   8   4

____ + ____ = ____
____ + ____ = ____
____ − ____ = ____
____ − ____ = ____

 Explain how you know the facts in a fact family.

Solve each problem. Show your work.

❶ Write three related facts for the number sentence below.

$$12 - 5 = 7$$

_____ + _____ = _____

_____ + _____ = _____

_____ − _____ = _____

❷ Write three related facts for the number sentence below.

$$6 + 8 = 14$$

_____ + _____ = _____

_____ − _____ = _____

_____ − _____ = _____

Choose the correct answer for each problem.

❸ Which sentence is not part of the fact family for 4, 6, and 10?

a) $10 + 4 = 14$

b) $4 + 6 = 10$

c) $10 - 4 = 6$

d) $6 + 4 = 10$

$$5 + 9 = 14$$
$$9 + 5 = 14$$
$$14 - 5 = 9$$
$$\underline{\quad} - \underline{\quad} = \underline{\quad}$$

❹ Which number sentence completes the fact family above?

a) $9 + 5 = 14$

b) $14 - 9 = 5$

c) $15 - 4 = 9$

d) $19 - 5 = 14$

Unit 2
Write a Number Sentence

Standard

Operations & Algebraic Thinking

Represent and solve problems involving addition and subtraction.

2.OA.1. Use addition and subtraction within 100 to solve one- and two-step word problems involving situations of adding to, taking from, putting together, taking apart, and comparing, with unknowns in all positions, e.g., by using drawings and equations with a symbol for the unknown number to represent the problem.

Model the Skill

Hand out 10 counters of one color and 10 counters of another color to student pairs. Then write the following problem on the board.

There are 5 red apples and 3 green apples. How many apples are there in all?

◆ Invite a student to read aloud the problem. **Say:** *You can show a math story problem with objects or pictures. Today we will use counters. How many counters should we show to represent the red apples? (5) How many counters should we show to represent the green apples? (3)* Have students use different colors for different color apples.

◆ **Ask:** *What is the question asking us to find out? (how many apples there are all together) Do you add the groups or compare the groups to find out? (add)* Observe as students put the groups together to find the total number of apples. **Ask:** *What number sentence shows this problem? (5 + 3 = 8)* Observe as students write a number sentence for the problem. Guide them to write the symbols in the circles and the numbers on the lines.

◆ Assign students the appropriate practice page(s) to support their understanding of the skill.

Assess the Skill

Use the following problems to pre-/post-assess students' understanding of the skill.

There are 6 green apples and 3 yellow apples. How many apples do we have in all?

We eat 2 apples. How many apples are left?

Name _____

Solve with counters. Complete each number sentence.

1 Max has 4 red cars and 5 green cars. How many cars does Max have all together?

___ ◯ ___ ◯ ___

2 Emma has 9 apples. She uses 8 to make a pie. How many apples does she have left?

___ ◯ ___ ◯ ___

3 Jordan has 7 cherries. He has two bowls. He puts 4 into one bowl. How many cherries are in the other bowl?

___ ◯ ___ ◯ ___

4 Will picks 8 plums. Jess gives Will 4 more plums. Then Will eats 2 plums. How many plums does Will have now?

___ ◯ ___ ◯ ___

___ ◯ ___ ◯ ___

☆ **Look at the page. In which problem did you add and subtract?** _____

Name _____

Draw a picture to solve. Write the number sentence.

1 Terry has 12 beads.
She used 8 on a necklace.
How many beads does she
have left?

● ● ● ● ● ● ● ● ● ● ● ● ●

___ ◯ ___ ◯ ___

2 Lucas made 9 drawings.
Then he made 7 more drawings.
How many drawings did he
make in all?

___ ◯ ___ ◯ ___

3 Ms. Smith bought 16 tomatoes.
She put 5 in a salad. How
many tomatoes does she
have left?

___ ◯ ___ ◯ ___

4 Sam put 13 flowers in a
vase. He put 7 yellow flowers
in the vase. The rest are
red. How many red flowers
are in the vase?

___ ◯ ___ ◯ ___

5 Frances bought 11 pencils.
Then she buys 8 more. She gives
2 pencils to Sally. How many
pencils does she have now?

___ ◯ ___ ◯ ___

___ ◯ ___ ◯ ___

6 Joe has 14 green stamps
and 3 blue stamps. He uses
4 stamps. How many stamps
does he have left?

___ ◯ ___ ◯ ___

___ ◯ ___ ◯ ___

☆ **Tell how you know what number sentence to write.**

Name _____

Write a number sentence for each word problem. Then solve.

1 Dia has 14 skirts. She puts 6 in a suitcase. The rest are in her closet. How many skirts are in her closet?

___ ◯ ___ ◯ ___

2 Mr. Ramos has 8 ties with stripes and 5 ties with dots. How many ties does he have in all?

___ ◯ ___ ◯ ___

3 Charlie has 9 pairs of socks. His father buys him 6 more pairs of socks. How many pairs of socks does Charlie have now?

___ ◯ ___ ◯ ___

4 Alex has 10 caps. Cassie has 8 caps. How many more caps does Alex have?

___ ◯ ___ ◯ ___

5 Mr. Thorn bought 6 blueberry muffins and 8 banana muffins. His children ate 3 muffins. How many muffins are left?

___ ◯ ___ ◯ ___
___ ◯ ___ ◯ ___

6 There are 16 rolls on a plate. James ate 3 rolls. Then Isaac ate 2 rolls. How many rolls are left?

___ ◯ ___ ◯ ___
___ ◯ ___ ◯ ___

7 Mrs. Bella bought 7 pizzas with pepperoni. She bought 2 plain pizzas. Her students ate 4 of the pizzas. How many pizzas are left?

___ ◯ ___ ◯ ___
___ ◯ ___ ◯ ___

8 There were 20 crackers on a plate. Sam ate 6 crackers. Then Rachel ate 7. How many crackers are still on the plate?

___ ◯ ___ ◯ ___
___ ◯ ___ ◯ ___

☆ **Tell how you solved Problem 8.**

Write a number sentence to solve each problem.

1 Sonia and Jay have 15 peaches all together. Sonia has 7 peaches. How many peaches does Jay have?

_____

2 Wendy had 6 purple beads and 5 green beads. Then she used 3 beads on a bracelet. How many beads does Wendy have now?

_____

3 We have 16 tarts. We have 3 more tarts in the oven. How many tarts do we have in all?

_____

4 Caden bakes 5 blueberry muffins and 7 lemon muffins. Her family eats 8 muffins. How many muffins does she have now?

_____

Choose the correct answer for each problem.

5 Fred has 7 melons. He sells 3 at the market. Then he eats 1. How many melons does he have left?

a) 3

b) 4

c) 5

d) 2

6 The Tigers scored 9 goals in all. They scored 4 goals in the second half of the game. How many goals did they score in the first half?

a) 3

b) 4

c) 5

d) 2

Unit 3
Odd and Even

Standard

> **Operations & Algebraic Thinking**
>
> **Work with equal groups of objects to gain foundations for multiplication.**
>
> **2.OA.3.** Determine whether a group of objects (up to 20) has an odd or even number of members, e.g., by pairing objects or counting them by 2s; write an equation to express an even number as a sum of two equal addends.

Model the Skill

Hand out 10 connecting cubes to each student. Then draw or model a 5-cube train.

◆ **Ask:** *How many cubes are in this train?* (5) Have students each make a cube train using 5 cubes. **Ask:** *How many are in a pair?* (2) *To find out if your cube train has an odd or even number, you can pair the cubes.* Guide them to break the train into pairs of cubes. To reinforce the cube pairs, have students draw a vertical line between every two cubes. **Ask:** *Does every cube have a partner?* (no) *If there is a cube without a partner, it means that the number of cubes is odd.*

◆ **Say:** *Add 1 cube to the 5-cube train to make 6. Then break the train into cube pairs.* Help students create their train and break it into 3 cube pairs. **Ask:** *Does every cube have a partner?* (yes) *If every cube has a partner, it means that the number of cubes is even.*

◆ Assign students the appropriate practice page(s) to support their understanding of the skill. Remind students to make pairs and that an item without a partner means that the number is odd.

Assess the Skill

Use the following problems to pre-/post-assess students' understanding of the skill.

Tell whether the number is odd or even.

2, 3, 5, 6, 7, 9, 12, 15, 16, 19, 20

Name _____

Circle whether the number of items is odd or even.

1

odd   even

2

odd   even

3

odd   even

4

odd   even

5

odd   even

6

odd   even

7

_____ + _____ = _____

odd   even

8

_____ + _____ = _____

odd   even

 Tell how you know an amount is even.

Name _____

Circle pairs. Then write whether the amount is odd or even.

❶

_____

❷

_____

❸

_____

❹

_____

❺

_____

❻

_____

Write whether the sum is odd or even.

❼ 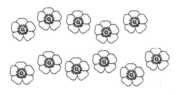 _____ + _____ = _____

_____

❽ _____ + _____ = _____

_____

❾ _____ + _____ = _____

_____

❿  _____ + _____ = _____

_____

☆ **Tell how you know whether to write *odd* or *even*.**

●●○

Circle pairs. Then write whether the amount is odd or even.

❶

_____

❷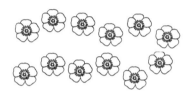

_____

❸

_____

❹

_____

❺

_____

❻

_____

Write whether the sum is odd or even.

❼

_____ + _____ = _____

_____

❽

_____ + _____ = _____

_____

❾

_____ + _____ = _____

_____

❿

_____ + _____ = _____

_____

⓫

_____ + _____ = _____

_____

⓬

_____ + _____ = _____

_____

 Explain why the sums of equal groups are always even.

Name _____

**For each word problem, write and solve an equation.
Then write whether the number is odd or even.**

1 Bill has 7 plums and Ava has 9 plums. In all, do they have an odd or an even number of plums?

_____

2 Alex and Soren each have 9 dollars. All together, do they have an odd or an even number of dollars?

_____

3 We have 14 rolls. We have 5 more rolls in the oven. Do we have an odd or an even number of rolls in all?

_____

4 I have 5 red pencils, 3 blue pencils, and 7 yellow pencils. Do I have an odd or even number of pencils?

_____

Choose the correct answer for each problem.

5 Jen has 8 tickets. She buys 3 more tickets. Then she gives 1 ticket away. Does she have an odd or even number of tickets left?

a) 9, odd

b) 10, even

c) 11, odd

d) 12, even

6 Oscar has 14 fish. He buys 5 more fish. Then he gives 3 fish away. Does he have an odd or even number of fish now?

a) 22, even

b) 19, odd

c) 16, even

d) 6, even

Unit 4
Add Equal Groups

Standard

Operations & Algebraic Thinking

Work with equal groups of objects to gain foundations for multiplication.

2.OA.4. Use addition to find the total number of objects arranged in rectangular arrays with up to 5 rows and up to 5 columns; write an equation to express the total as a sum of equal addends.

Model the Skill

Hand out counters and draw a 3 x 2 array of circles on the board or on paper.

◆ **Say:** *The arrangement of objects in equal rows and columns is called an array. This is an array. An array has columns and rows. A column goes up and down while a row goes across.* Have students copy the array onto paper. **Say:** *Place a counter on each circle in the first column (or color the column in). How many counters did you use?* (2) Point out the number 3 at the bottom of the column. **Say:** *Now place a counter on each circle in the next column. How many counters did you use?* (3) *Look at the addition sentence below the counters. What is 3 plus 3?* (6) Observe as students write the sum of 6.

◆ Draw a 3 x 3 array. **Say:** *The arrangement of objects in equal rows and columns is called an array. How is the first array different from this array?* (Possible answer: It has one more column.) Place a counter on each circle in a column, or color in, and count how many counters are in each column. Check that students are placing counters in columns rather than rows. **Ask:** *What is 3 plus 3 plus 3?* (9) *Is that the total number of counters you have used?* (yes)

◆ Assign students the appropriate practice page(s) to support their understanding of the skill. **Say:** *You can use addition to find the total number of items in an array.*

Assess the Skill

Use the following problems to pre-/post-assess students' understanding of the skill.

Ask students to use repeated addition to find the sum total for each array.

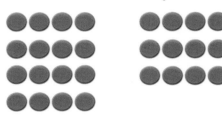

Name _____

Add to find the total.

❶

2 + 2 = _____

❷

2 + 2 + 2 = _____

---

❸

2 + 2 + 2 + 2 = _____

❹

2 + 2 + 2 + 2 + 2 = _____

---

❺

3 + 3 + 3 + 3 = _____

❻

4 + 4 + 4 = _____

☆ **Tell how you can use skip counting to find the total.**

Name _____

Write the number of items in each column. Add to find the total.

1

2 + 2 + 2 = ____

2

___ + ___ = ___

3

___ + ___ + ___ = ___

4

___ + ___ + ___ + ___ + ___ = ___

5

___ + ___ + ___ + ___ = ___

6

___ + ___ + ___ + ___ = ___

7

___ + ___ + ___ = ___

8

___ + ___ + ___ + ___ + ___ = ___

 How do your answers for 6 and 7 compare? Tell how the arrays are alike and different.

Name _____

Write a number sentence for each array.

1 ◯ ◯
◯ ◯
◯ ◯

___ + ___ = ___

2 ◯ ◯ ◯
◯ ◯ ◯

___ + ___ + ___ = ___

3 ◯ ◯ ◯ ◯
◯ ◯ ◯ ◯

___ + ___ +___ + ___ = ___

4

___ + ___ + ___ = ___

5

___ + ___ + ___ = ___

6

___ + ___ + ___ + ___ = ___

7

_____

8

_____

9

_____

10

_____

☆ **Tell how skip counting can help you find the total number in an array.**

Write a number sentence for each array.

1 Teddy, Meta, and Henry each have 5 dollars. How many dollars do they have in all?

_____

2 The grocer has 3 rows of peppers. Each row has 8 peppers. How many peppers does the grocer have in all?

_____

3 The vegetable garden has 4 rows of tomato plants. Each row has 9 tomato plants. How many tomato plants are there in all?

_____

4 Annabelle has 5 boxes of raisins. Each box has 10 raisins. How many raisins does she have in all?

_____

Choose the correct answer for each problem.

5 There are 3 cartons of eggs. Each carton has 6 eggs. How many eggs are there in all?

a) 9

b) 12

c) 18

d) 36

6 There are 6 bags of apples. There are 7 apples in each bag. How many apples are there in all?

a) 13

b) 26

c) 42

d) 35

Unit 5
Understand Place Value

Standard

Number & Operations in Base Ten

Understand place value.

2.NBT.1. Understand that the three digits of a three-digit number represent amounts of hundreds, tens, and ones; e.g., 706 equals 7 hundreds, 0 tens, and 6 ones. Understand the following as special cases:

2.NBT.1.a. 100 can be thought of as a bundle of ten tens—called a "hundred."

2.NBT.1.b. The numbers 100, 200, 300, 400, 500, 600, 700, 800, 900 refer to one, two, three, four, five, six, seven, eight, or nine hundreds (and 0 tens and 0 ones).

Model the Skill

Hand out base-ten blocks.

◆ **Say:** *A number can be shown with base-ten blocks.* Review with students that 10 ones equal 1 ten. Then show 10 tens. **Ask:** *What number is shown with 10 tens?* (100) Have students count the rods by tens up to 100. **Say:** *Use blocks to show the number 123. How many hundreds do you use?* (1) *How many tens?* (2) *How many ones?* (3)

◆ **Ask:** *How do you show the number 200?* (Possible answer: with 2 hundreds) *Why are there no tens or ones shown in the chart?* Write the number 300 on the board. **Ask:** *Why are there zeros in the tens and ones places?* (Possible answer: because there are no tens or ones in the number 300) *Even though only 3 hundreds are needed to show the number, you need to write the zeros in the tens and ones places to indicate that no tens and ones are needed.*

◆ Assign students the appropriate practice page(s) to support their understanding of the skill.

Assess the Skill

Use the following problem to pre-/post-assess students' understanding of the skill.

	hundreds	tens	ones
407			

Name _____

Show each number with base-ten blocks. Write the number of hundreds, tens, and ones.

① 234

hundreds	tens	ones

② 300

hundreds	tens	ones

③ 209

hundreds	tens	ones

④ 145

hundreds	tens	ones

⑤ 266

hundreds	tens	ones

⑥ 159

hundreds	tens	ones

☆ **Tell how you know what to write in each column.**

27

Name _____

Write the number in the correct places of the chart.

1

350

hundreds	tens	ones

2

400

hundreds	tens	ones

3

308

hundreds	tens	ones

4

731

hundreds	tens	ones

5

221

hundreds	tens	ones

6

298

hundreds	tens	ones

7

469

hundreds	tens	ones

☆ **Tell how you know what number to write in each place.**

 Unit 5 • Common Core Mathematics Grade 2 • ©2012 Newmark Learning, LL

Name _____

Write the number of hundreds, tens, and ones. Then write the number.

1

hundreds	tens	ones
3	1	8

____ hundreds
____ tens
____ ones =
____

2

hundreds	tens	ones
4	5	2

____ hundreds
____ tens
____ ones =
____

3

hundreds	tens	ones
7	3	0

____ hundreds
____ tens
____ ones =
____

4

hundreds	tens	ones
2	2	7

____ hundreds
____ tens
____ ones =
____

5

hundreds	tens	ones
5	2	4

____ hundreds
____ tens
____ ones =
____

6

hundreds	tens	ones
9	3	0

____ hundreds
____ tens
____ ones =
____

7

hundreds	tens	ones
1	3	6

____ hundreds
____ tens
____ ones =
____

8

hundreds	tens	ones
3	2	8

____ hundreds
____ tens
____ ones =
____

9

hundreds	tens	ones
4	1	1

____ hundreds
____ tens
____ ones =
____

10

hundreds	tens	ones
7	0	0

____ hundreds
____ tens
____ ones =
____

11

hundreds	tens	ones
5	4	0

____ hundreds
____ tens
____ ones =
____

12

hundreds	tens	ones
6	0	8

____ hundreds
____ tens
____ ones =
____

 Tell how you know the number of tens.

●●● 29

Write the number of hundreds, tens, and ones.

1 The number 328 has

_____ hundreds

_____ tens

_____ ones

2 The number 340 has

_____ hundreds

_____ tens

_____ ones

3 The number 161 has

_____ hundreds

_____ tens

_____ ones

4 The number 509 has

_____ hundreds

_____ tens

_____ ones

5 The number 442 has

a) Four ones, four hundreds, and two tens

b) Four hundreds, two tens, and two ones

c) Four hundreds, four tens, and ten ones

d) Four hundreds, four tens, and two ones

6 The number 671 has

a) Six hundreds, seven ones, and two tens

b) Seven hundreds, two tens, and six ones

c) Six hundreds, seven tens, and one one

d) Six hundreds, seven tens, and zero ones

Unit 6

Count, Read, and Write Numbers to 1,000

Standard

> **Number & Operations in Base Ten**
>
> **Understand place value.**
>
> **2.NBT.2.** Count within 1,000; skip count by 5s, 10s, and 100s.
>
> **2.NBT.3.** Read and write numbers to 1,000 using base-ten numerals, number names, and expanded form.

Model the Skill

Hand out base-ten blocks and write the following on the board.

hundreds	tens	ones
2	6	5

265

_____ hundreds _____ tens _____ ones

Expanded Form: _____00 + _____0 + _____

◆ **Say:** _Today we will show the same number in different ways._ Read aloud the number on the board. (two hundred sixty-five) Show that number with base-ten blocks. Have students build the number with blocks. Guide them to write the number of hundreds, tens, and ones.

◆ **Say:** _You can show a number as a sum of the hundreds, tens, and ones. Look at the third column. What would you write to show the numeral for 2 hundreds?_ (a 2 in front of the zeros) _What do you need to write to show the numeral for 6 tens?_ (a 6 in front of the zero) _What do you need to write to show the numeral for 5 ones?_ (a 5) Refer to all three columns to show the same number in different ways.

◆ Assign students the appropriate practice pages to support their understanding of the skill. Remind them that they need to write the zeros for the hundreds as well as the tens in the expanded form.

Assess the Skill

Use the following problems to pre-/post-assess students' understanding of the skill.

Write each number in standard, expanded, and word form.

721   582   490   817   968   101

Count by 5s from 600 to 650.

Count by 10s from 800 to 900.

Name _____

Write the number in the place value chart in all forms.

1

hundreds	tens	ones
3	4	0

standard form: **340**

expanded form: _____00 + _____0 + _____

word form _____

2

hundreds	tens	ones
7	1	2

standard form: _____

expanded form: _____00 + _____0 + _____

word form: _____

3   **Skip count by 100s. Continue counting from the numbers given.**

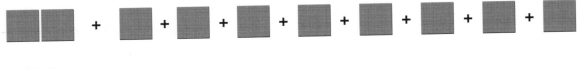

200,   _____ , _____ , _____ , _____ , _____ , _____ , _____ , _____

4   **Skip count by 10s.**

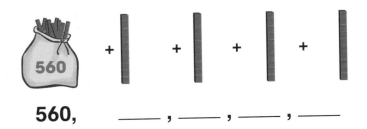

560,   _____ , _____ , _____ , _____

5   **Skip count by 5s.**

810   815   820   _____   _____   _____   840

☆   **Tell how you know how to read the number.**

Name _____

Write the number shown in three different ways. Use base-ten blocks if you wish.

❶

hundreds	tens	ones
1	9	0

standard form: _____

expanded form: ____00 + ____0 + ____

word form: _____

---

❷ standard form: **249**   expanded form: ____ + ____ + ____

word form: _____

---

❸ standard form: _____   expanded form: **600 + 20 + 8**

word form: _____

---

❹ standard form: _____   expanded form: ____ + ____ + ____

word form: **eight hundred thirty-four**

---

❺ Skip count by 100s.

300 400 500 ___ ___ ___ ___

❻ Skip count by 10s.

460 470 480 ___ ___ ___ ___

---

❼ Skip count by 5s.

510 515 520 ___ ___ ___ ___

❽ Look for a skip-counting pattern. Write the missing numbers.

630, 640, 650, ____, ____, ____

⭐ **Tell how you know the different ways to write a number.**

Name _____

Fill in the blanks for each problem.

❶ standard form: **732**   expanded form: _____ **+** _____ **+** _____

     word form: _____

❷ standard form: _____   expanded form: **500 + 90 + 4**

     word form: _____

❸ standard form: _____   expanded form: _____ **+** _____ **+** _____

     word form: **six hundred nine**

❹ standard form: **813**   expanded form: _____ **+** _____ **+** _____

     word form: _____

Look for a skip-counting pattern. Write the missing numbers.

❺
325, 330, 335, _____, _____, _____

❻
500, 600, 700, _____, _____, _____

❼
740, 750, 760, _____, _____, _____

❽
445, 450, 455, _____, _____, _____

❾
450, 550, 650, _____, _____, _____

❿
975, 980, 985, _____, _____, _____

☆ **Tell how you know which number comes next.**

Write the number in the missing forms.

1 standard form: **768**

expanded form: _____

word form: _____

2 standard form: _____

expanded form: _____

word form: **four hundred seventy**

Look for a skip-counting pattern. Write the missing numbers.

3 630, 640, 650, _____, _____, _____

4 260, 270, 280, _____, _____, _____

5 945, 950, 955, _____, _____, _____

6 705, 710, 715, _____, _____, _____

7 420, 430, 440, _____, _____, _____

8 685, 690, 695, _____, _____, _____

9 825, 830, 835, _____, _____, _____

10 320, 330, 340, _____, _____, _____

Choose the correct answer for each problem.

11 Look at the pattern below. What number comes next?

805, 810, 815, _____

a) 825

b) 816

c) 915

d) 820

12 Look at the pattern below. What number comes next?

780, 790, 800, _____

a) 880

b) 810

c) 801

d) 900

Unit 7
Compare Numbers

Standard

Number & Operations in Base Ten

Understand place value.

2.NBT.4. Compare two three-digit numbers based on meanings of the hundreds, tens, and ones digits, using >, =, and < symbols to record the results of comparisons.

Model the Skill

Hand out base-ten blocks and write the following values on the board.

123   **132**

◆ **Say:** *Let's compare these two numbers. Use your base-ten blocks to make each number.* Show the numbers in separate locations. **Say:** *First, you need to compare the greatest place—the hundreds. Does one number have more hundreds than the other?* (no) *Compare the next place—the tens. Does one number have more tens than the other?* (Yes, 132 has more tens.) Point out that there is no need to compare the ones because they have already determined which number is greater. **Say:** *132 is greater than 123 and 123 is less than 132.* Then practice another example with students.

◆ **Say:** *Now compare 205 and 210. This group of 205 must be greater because it has more blocks. Am I right?* (no) *Why not?* (Possible answer: Having more blocks doesn't mean the number is greater; you have to look at the values of the blocks.) *1 ten is equal to 10 ones. If you showed 210 with 2 hundreds and 10 ones, there would be more blocks. 210 is greater than 205 and 205 is less than 210.*

◆ Assign students the appropriate activity page(s) to support their understanding of the skill.

Assess the Skill

Use the following problems to pre-/post-assess students' understanding of the skill.

Use place value to compare numbers. Write <, >, =.

354 ◯ 364   732 ◯ 732   923 ◯ 629

Put these numbers in order of least to greatest in value.

402   539   413   699   960   331

Compare these numbers. Circle the number that is greater. Then write the correct sign for each problem.

> > is greater than
> < is less than
> = is equal to

①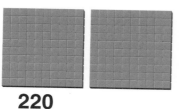

145   154

②

220   215

③

237   234

④

140   228

⑤

hundreds	tens	ones
2	3	0

230

hundreds	tens	ones
2	0	3

203

⑥

hundreds	tens	ones
2	2	7

227

hundreds	tens	ones
1	7	2

172

 Tell how you know which number is greater.

Name _____

Compare each set of numbers. Circle the true statement. Then write the correct sign for each problem.

> is greater than

< is less than

= is equal to

1

215 is greater than 213.
215 is less than 213.
215 is equal to 213.

215 ◯ 213

2

hundreds	tens	ones
8	4	5
8	4	1

845 is greater than 841.
845 is less than 841.
845 is equal to 841.

845 ◯ 841

3

hundreds	tens	ones
7	3	9
7	7	9

739 is greater than 779.
739 is less than 779.
739 is equal to 779.

739 ◯ 779

Write the correct sign for each problem.

4 347 ◯ 374   **5** 681 ◯ 918   **6** 720 ◯ 702

7 457 ◯ 475   **8** 167 ◯ 171   **9** 297 ◯ 297

☆ Tell how you know when a number is less than the other number.

  Unit 7 • Common Core Mathematics Grade 2 • ©2012 Newmark Learning, LL

Name _____

Compare each set of numbers.
Then write the correct sign for each problem.

>	is greater than
<	is less than
=	is equal to

①

hundreds	tens	ones
6	3	2
6	4	1

632 ◯ 641

②

hundreds	tens	ones
7	2	1
4	3	0

721 ◯ 430

③   631 ◯ 913

④   802 ◯ 802

⑤   654 ◯ 954

⑥   179 ◯ 181

⑦   218 ◯ 128

⑧   709 ◯ 706

⑨   454 ◯ 462

⑩   308 ◯ 311

⑪   577 ◯ 574

⑫   356 ◯ 356

 Tell how you know which symbol to write.

Name _____

Solve.

1 Sally has 231 baseball cards. Matt has 229 baseball cards. Who has a greater number of baseball cards?

2 The cheese store sold 329 pounds of cheese on Monday. They sold 259 pounds of cheese on Tuesday. On which day did they sell more cheese?

Circle the correct answer for each problem.

3 Claudia has 623 beads in her collection. Liv has 632 beads. Who has more beads?

a) Claudia

b) Liv

c) They have an equal number of beads

4 Claudia has 623 beads in her collection. If she trades 5 beads to Liv for 5 beads, how many will Claudia have?

a) 618 beads

b) 613 beads

c) 623 beads

d) 628 beads

Unit 8
Use Strategies to Add

Standard

Number & Operations in Base Ten

Use place value understanding and properties of operations to add and subtract.

2.NBT.5. Fluently add and subtract within 100 using strategies based on place value, properties of operations, and/or the relationship between addition and subtraction.

Model the Skill

Write the following number line and addition sentence on the board.

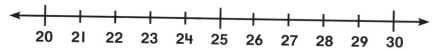

20 21 22 23 24 25 26 27 28 29 30

$$25 + 2 =$$

◆ **Say:** *You can use different strategies to add. You can count on by 1, 2, or 3. A number line can help you.* Have students use the number line to count on and find the addend. **Say:** *You can draw two jumps to count 2 past 25.* Show students how to draw a curved line from 25 to 26 and then from 26 to 27 to show two jumps. **Ask:** *On what number did you land?* (27) Then write another problem on the board.

$$27 + 3 =$$

◆ Have students look at the problem. **Ask:** *What number will you circle on the number line?* (27) *How many jumps will you make?* (3) *What is 27 plus 3?* (30)

◆ Assign students the appropriate activity page(s) to support their understanding of the skill. Remind students using number lines to circle the first addend and draw jumps to the right on the number line equal to the second addend.

Assess the Skill

Use the following problems to pre-/post-assess students' understanding of the skill.

Solve.

$$16 + 7 = \qquad 37 + 8 = \qquad 41 + 7 + 19 = \qquad 83 + 6 + 14 =$$

Name _____

Count on to add. Write each sum.

❶
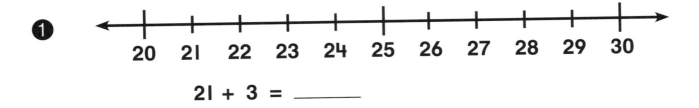

21 + 3 = _____

❷
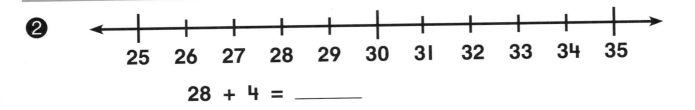

28 + 4 = _____

❸
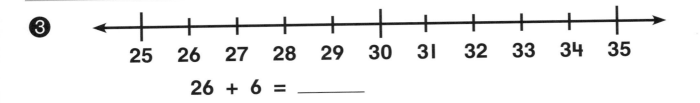

26 + 6 = _____

❹
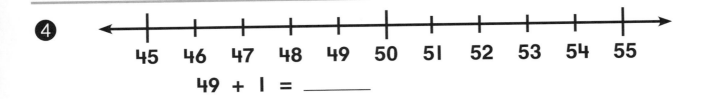

49 + 1 = _____

❺
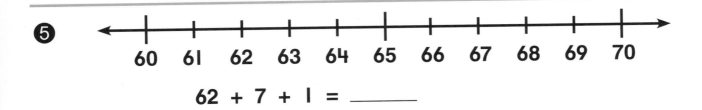

62 + 7 + 1 = _____

❻
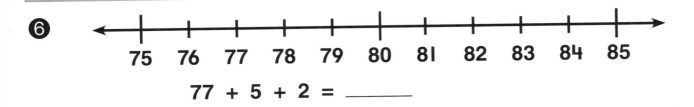

77 + 5 + 2 = _____

☆ **Tell how you can use a number line to add.**

   Unit 8 • Common Core Mathematics Grade 2 • ©2012 Newmark Learning, LLC

Name _____

Add in any order. Show your thinking. Write each sum.

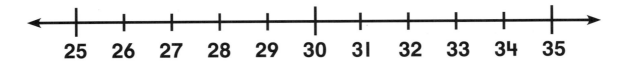

❶ 3 + 28 = _____

❷ 26 + 7 = _____

❸ 8 + 84 = _____

❹ 2 + 53 = _____

Circle the numbers you will add first. Show your work. Write the sum.

❺ 32 + 15 = _____

❻ 7 + 41 = _____

❼ 33 + 15 + 5 = _____

❽ 7 + 42 + 11 = _____

 Tell how you solved the problem.

Name _____

Find each sum.

1   41 + 9 = _____

2   9 + 30 = _____

3   8 + 23 = _____

4   16 + 5 = _____

5   8 + 34 = _____

6   9 + 28 = _____

7   56 + 5 = _____

8   9 + 44 = _____

9   7 + 63 = _____

10   76 + 5 + 3 = _____

11   5 + 37 + 4 = _____

12   8 + 23 + 14 = _____

☆   **Tell how you added the three numbers in Problem 11.**

Solve.

1 Jake has 12 cards and Mary has 8 cards. How many cards do they have in all?

2 Catherine has 6 red socks, 18 white socks, and 14 purple socks. How many socks does she have in all?

3 Maya has 18 pennies. Her brother gives her 2 more pennies. Then her Dad gives her 6 pennies. How many pennies does she have in all?

4 Rob made 37 cupcakes for the party. Jan brought 8 cupcakes. Eddie brought 2 more. How many cupcakes did they have at the party?

Choose the correct answer for each problem.

5 Jill read 13 pages of her book on Monday. She read 5 more pages on Tuesday. Then she read 7 pages on Friday. How many pages did she read in all?

a) $13 + 5 + 7 = 27$

b) $13 + 5 + 7 = 24$

c) $13 + 5 + 7 = 25$

d) $13 + 5 + 7 = 28$

6 Ariba and Spencer both have 8 grapes. Then Ariba's mom gives them each 6 more. How many grapes do they have in all?

a) $8 + 8 + 6 = 22$

b) $8 + 6 + 6 = 20$

c) $8 + 8 + 6 + 6 = 26$

d) $8 + 8 + 6 + 6 = 28$

Unit 9
Add Two-Digit Numbers

Standard

Number & Operations in Base Ten

Use place value understanding and properties of operations to add and subtract.

2.NBT.6. Add up to four two-digit numbers using strategies based on place value and properties of operations.

Model the Skill

Hand out base-ten blocks and write the following problem on the board.

$$41 + 26 =$$

◆ **Say:** *Let's add these amounts by first joining the tens. How many tens are there in all? (6) What is 40 + 20?* Have students write the sum. (60) **Say:** *Let's add the ones. How many ones are there in all? (7) What is 1 + 6?* Have students write the sum. (7) **Say:** *Now add the sum of the tens and the sum of the ones. What is the total sum? (67)* Then write another problem on the board.

$$63 + 15 =$$

◆ **Ask:** *How would you show how to add the tens? (60 + 10) How would you show how to add the ones? (3 + 5)* Have students show the numbers with blocks, add the tens (70), and add the ones (8). **Say:** *Once you have found the sum of the tens and ones, what do you do? (Add the tens and ones together.)* Write the partial sums and calculate the total sum. (78)

◆ Assign students the appropriate practice page(s) to support their understanding of the skill. Encourage students to model the addends, adding the tens, adding the ones, and writing the partial sums if necessary.

Assess the Skill

Use the following problems to pre-/post-assess students' understanding of the skill.

Solve.

$22 + 47 =$ $53 + 14 =$ $67 + 12 =$ $81 + 16 =$ $34 + 55 =$

$23 + 47 =$ $63 + 18 =$ $58 + 14 =$ $72 + 19 =$ $34 + 58 =$

$23 + 47 + 11 =$ $63 + 12 + 20 =$ $48 + 14 + 32 + 25 =$

Name _____

Find each sum.

❶  25 + 40 =

	tens	ones
	2	5
+	4	0

❷ 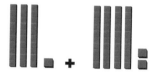 31 + 42 =

	tens	ones
	3	1
+	4	2

❸  46 + 13 =

	tens	ones
+		

❹  14 + 35 =

	tens	ones
+		

❺ 46 + 14 + 30 =

	tens	ones
+		

❻ 66 + 12 + 21 =

	tens	ones
+		

 Tell how you add two-digit numbers.

Name _____

Find each sum.

1   54 + 18

tens	ones
5	4
1	8

2   29 + 37

tens	ones
2	9
3	7

3   25 + 73

4   54 + 26

5   46 + 35

6   22 + 69

7   53 + 18 + 32

8   48 + 17 + 25

    Tell how you know when you have to regroup.

Name _____

Find the sum.

1   12 + 48 =

2   55 + 16 =

3   34 + 22 + 18 =

4   41 + 13 + 27 =

5   54 + 16 + 24 =

6   42 + 13 + 45 =

7   21 + 57 + 19 =

8   26 + 43 + 11 =

9   43 + 27 + 18 =

10   37 + 18 + 25 =

   Tell how you found the sum.

Solve.

1 Last season Nina scored 17 goals. This season she scored 14 goals. How many goals has she scored in all?

2 Alex picked 41 apples. Sam picked 29 apples. How many apples did they pick in all?

Choose the correct answer for each problem.

3 Ryan has 14 marbles. Jaya has 25 marbles. Graham has 36 marbles. How many marbles do they have in all?

a) 74

b) 85

c) 65

d) 75

4 Ms. Eves has 22 students in her class. Mr. Jacobs has 25 students. Ms. Repo has 29 students. How many students are in all three classes?

a) 66

b) 76

c) 78

d) 86

Unit 10
One Hundred More, One Hundred Less

Standard

Number & Operations in Base Ten

Use place value understanding and properties of operations to add and subtract.

2.NBT.8. Mentally add 10 or 100 to a given number 100–900, and mentally subtract 10 or 100 from a given number 100–900.

Model the Skill

Hand out base-ten blocks and write the following problems on the board:

$$51 + 10 =$$

$$51 + 100 =$$

◆ Have students show 51 with their blocks. **Say:** *Add a ten. Now how many do you have?* (61) Have students return their blocks to showing 51. **Say:** *Now take away a ten. What number do the blocks show?* (41) Continue to practice adding and subtracting 10 from a given number.

◆ **Say:** *Today you will be adding one hundred to some numbers. What number are you adding to 51?* (100) Observe as students add a hundred to their blocks. **Ask:** *What number do your blocks show now?* (151) Have students write the sum. Then guide students to complete the other problems on the board.

◆ **Ask:** *What is the first addend in 162 + 100?* (162) Observe as students show 1 hundred, 6 tens, and 2 ones blocks. **Ask:** *What are you adding to 162?* (100) *How do you show that?* (Possible answer: I place another hundred block.) *What is the sum of 162 and 100?* (262)

◆ Assign students the appropriate practice page(s) to support their understanding of the skill.

Assess the Skill

Use the following problems to pre-/post-assess students' understanding of the skill.

$24 + 100 =$ _____ $350 - 100 =$ _____

$168 + 100 =$ _____ $764 - 100 =$ _____

$240 + 100 =$ _____ $509 - 100 =$ _____

Name _____

Show the addition with blocks. Write the sum.

❶

62 + 100 = _____

 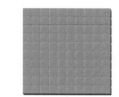

❷

135 + 100 = _____

❸

214 + 100 = _____

 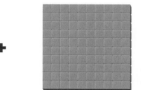

❹

250 – 100 = _____

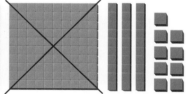

❺

239 – 100 = _____

 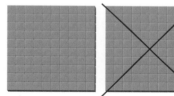 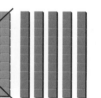

❻

302 – 100 = _____

 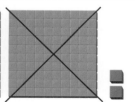

 Tell which place changes when you add or subtract one hundred.

●○○

Name _____

Find the sum or the difference for each problem.

1   133 + 100 = ____

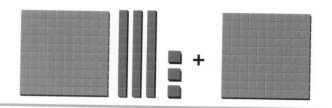

2   278 + 100 = ____

3   360 + 100 = ____

4   201 + 100 = ____

5   275 − 100 = ____

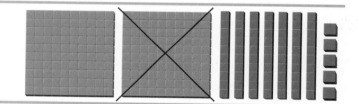

6   146 − 100 = ____

7   679 − 100 = ____

8   705 − 100 = ____

   Tell how you subtract one hundred.

Name _____

Find each sum or difference.

1   547 + 100

2   874 – 100

3   358 – 100

4   441 + 100

5   639 + 100

6   923 – 100

7   688 + 100

8   140 – 100

9   753 – 100

10   601 + 100

☆   **Tell how you subtract 100 from 601.**

Solve.

1 The florist has 260 roses. He sells 100 roses. How many roses does he have left?

2 The baker makes 435 cookies. Then she bakes another batch of 100 cookies. How many cookies does she have in all?

Circle the correct answer for each problem.

3 Allison has 452 bees in her beehive. She gets 100 more bees. How many bees does she have in all?

a) 352 bees

b) 252 bees

c) 552 bees

d) 442 bees

4 Jane has 150 dollars. She spends 100 dollars. How many dollars does she have left?

a) 250 dollars

b) 150 dollars

c) 140 dollars

d) 50 dollars

 Tell how the first number compares with the answer.

Unit 11
Add Three-Digit Numbers

Standard

Number & Operations in Base Ten

Use place value understanding and properties of operations to add and subtract.

2.NBT.7. Add and subtract within 1,000, using concrete models or drawings and strategies based on place value, properties of operations, and/or the relationship between addition and subtraction; relate the strategy to a written method. Understand that in adding or subtracting three-digit numbers, one adds or subtracts hundreds and hundreds, tens and tens, ones and ones; and sometimes it is necessary to compose or decompose tens or hundreds.

Model the Skill

Hand out base-ten blocks and write the following problem on the board.

$$162 + 135 =$$

◆ **Say:** *Today we are going to add 2 three-digit numbers. A three-digit number has hundreds, tens, and ones.* Ask students to show the two addends with blocks. Observe as students build the numbers. **Say:** *Add the hundreds together. How many are there?* (2) *That means that 100 plus 100 is 200.* Guide students to write that sum on a piece on paper. **Ask:** *What amounts are we going to add next?* (60 and 30) Guide students to join the tens and write 9 tens as 90. Then guide students to join the ones and write the sum on the page. (7) **Say:** *Now add the three place-value sums. What is the total sum?* (297) Then write another problem on the board.

$$237 + 125 =$$

◆ Have students show the addends in 237 + 125. **Ask:** *What is the first step?* (Add the hundreds.) *What is the sum?* (300) *What is the second step?* (Add the tens.) *What is the sum?* (50) *What is the third step?* (Add the ones.) *What is the sum?* (12) Have students add the numbers together to find the total. (362)

◆ Assign the appropriate practice page(s) to support each student's understanding of the skill.

Assess the Skill

Use the following problems to pre-/post-assess students' understanding of the skill.

$$143 + 126 =$$

$$234 + 145 =$$

$$343 + 227 =$$

$$573 + 140 =$$

$$422 + 369 =$$

Name _____

Show each addend with blocks. Add each place. Add the sums.

❶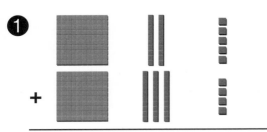

$125 + 134 =$

$100 + 100 =$ _____ ⇨ _____

$20 + 30 =$ _____ ⇨ _____

$5 + 4 =$ _____ ⇨ + _____

_____

❷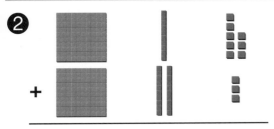

$118 + 123 =$

$100 + 100 =$ _____ ⇨ _____

$10 + 20 =$ _____ ⇨ _____

$8 + 3 =$ _____ ⇨ + _____

_____

❸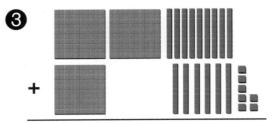

$290 + 167 =$

$200 + 100 =$ _____ ⇨ _____

$90 + 60 =$ _____ ⇨ _____

$0 + 7 =$ _____ ⇨ + _____

_____

❹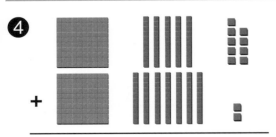

$159 + 172 =$

$100 + 100 =$ _____ ⇨ _____

$50 + 70 =$ _____ ⇨ _____

$9 + 2 =$ _____ ⇨ + _____

_____

❺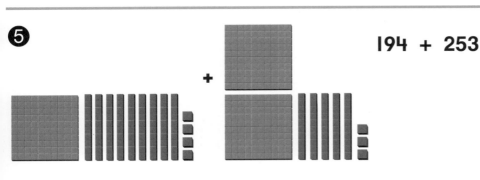

$194 + 253$

	hundreds	tens	ones
	1	9	4
+	2	5	3

 Tell how you found the sum.

Name _____

Find each sum.

❶

163 + 121

	hundreds	tens	ones
	1	6	3
+	1	2	1

❷

244 + 122

	hundreds	tens	ones
	2	4	4
+	1	2	2

❸ 343 + 203 =

	hundreds	tens	ones
+			

❹ 174 + 703 =

	hundreds	tens	ones
+			

❺ 516 + 382 =

	hundreds	tens	ones
+			

❻ 720 + 218 =

	hundreds	tens	ones
+			

❼ 674 + 241 =

	hundreds	tens	ones
+			

❽ 405 + 398 =

	hundreds	tens	ones
+			

  Tell how you add three-digit numbers.

● ● ○

Find each sum.

1

147 + 135

hundreds	tens	ones
1	4	7
+ 1	3	5

2   258 + 126 =

3   193 + 181 =

4   176 + 343 =

5   341 + 172 =

6   409 + 251 =

7   361 + 307 =

8   286 + 244 =

9   560 + 296 =

10   276 + 257 =

11   193 + 181 =

12   276 + 141 =

 Tell how you solved the problem.

Name _____

Solve.

1 Ann has 546 toothpicks. Jim has 325 toothpicks. How many toothpicks do they have in all?

2 Teddy has 433 paper clips. Sara has 292 paper clips. How many do they have in all?

Circle the correct answer for each problem.

3 Paige has 228 stamps. Ben has 143 stamps. How many stamps do they have in all?

a) 361 stamps

b) 371 stamps

c) 381 stamps

d) 261 stamps

4 Chase had 152 dollars in the bank. He puts 125 more dollars in the bank. How many dollars does he have in the bank now?

a) 250 dollars

b) 275 dollars

c) 277 dollars

d) 27 dollars

Unit 12
Use Strategies to Subtract

Standard

Number & Operations in Base Ten

Use place value understanding and properties of operations to add and subtract.

2.NBT.9. Explain why addition and subtraction strategies work, using place value and the properties of operations.

Model the Skill

Draw a number line on the board.

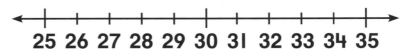

25 26 27 28 29 30 31 32 33 34 35

$$33 - 2 =$$

◆ **Say:** *You can use different strategies to subtract. You can count back by 1, 2, or 3 on a number line.* Have students look at the problem and circle the first number in the equation on the number line. **Say:** *You can draw two jumps to count back 2 from 33.* Show students how to draw a curved line from 33 to 32 and then to 31. Encourage students to count back aloud as they draw the jumps. **Ask:** *On what number did you land?* (31) Then write another problem on the board.

$$32 - 5 =$$

◆ Have students repeat with 32 – 5. Observe as students draw the jumps and count back aloud. **Ask:** *What are some other strategies we use when we subtract? Do we think about fact families? Do we use our understanding of place value?* Encourage students to explain their different methods for subtraction problem solving.

◆ Assign the appropriate practice page(s) to support each student's understanding of the skill.

Assess the Skill

Use the following problems to pre-/post-assess students' understanding of the skill.

$$13 - 7 =$$

$$23 - 7 =$$

$$36 - 4 =$$

$$47 - 9 =$$

Name _____

Count back to subtract. Write each difference.

❶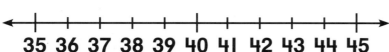

35 36 37 38 39 40 41 42 43 44 45

$42 - 3 =$ _____

❷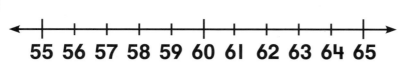

55 56 57 58 59 60 61 62 63 64 65

$65 - 1 =$ _____

❸ $11 - 4 =$ _____

Think: $4 + ? = 11$

❹ $14 - 8 =$ _____

Think: $8 + ? = 14$

❺

tens	ones
7	3
−	2

$73 - 2 =$ _____

❻

tens	ones
4	6
−	9

$46 - 9 =$ _____

 Tell how you use place value to subtract.

Unit 12 • Common Core Mathematics Grade 2 • ©2012 Newmark Learning, L

Name _____

Use a related addition fact to subtract.

1

$17 - 9 = $ _____

> Think:
> $9 + ? = 17$

2

$15 - 7 = $ _____

> Think:
> $7 + ? = 15$

3

$14 - 8 = $ _____

> Think:
> $8 + ? = 14$

4

$11 - 4 = $ _____

> Think:
> $4 + ? = 11$

5

$18 - 9 = $ _____

6

$16 - 8 = $ _____

7

$12 - 7 = $ _____

8

$13 - 6 = $ _____

 Tell how you can use an addition fact to help you subtract.

Name _____

Find each difference.

❶

49 − 4 =

tens	ones
−

❷

29 − 7 =

tens	ones
−

❸

38 − 6 =

tens	ones
−

❹ 75 − 4 =

❺ 84 − 7 =

❻ 67 − 6 =

❼ 60 − 4 =

❽ 42 − 8 =

❾ 37 − 5 =

❿ 26 − 9 =

⓫ 81 − 6 =

⓬ 53 − 7 =

☆ **Tell how you solved Problem 12.**

Solve.

1 Jude has 84 crayons. He gives 3 to Lily. How many crayons does Jude have left?

2 Maddie has 75 markers. She gives 7 to Carl. How many markers does Maddie have left?

Circle the correct answer for each problem.

3 Shen has 81 stamps. He uses 5 stamps to mail some letters. How many stamps does he have left?

a) 76 stamps

b) 86 stamps

c) 77 stamps

d) 31 stamps

4 Farah has 63 cents. She spends 8 cents. How many cents does she have left?

a) 71 cents

b) 66 cents

c) 55 cents

d) 23 cents

Unit 13
Subtract Two-Digit Numbers

Standard

Number & Operations in Base Ten

Use place value understanding and properties of operations to add and subtract.

2.NBT.5. Fluently add and subtract within 100 using strategies based on place value, properties of operations, and/or the relationship between addition and subtraction.

2.NBT.7. Add and subtract within 1,000, using concrete models or drawings and strategies based on place value, properties of operations, and/or the relationship between addition and subtraction; relate the strategy to a written method. Understand that in adding or subtracting three-digit numbers, one adds or subtracts hundreds and hundreds, tens and tens, ones and ones; and sometimes it is necessary to compose or decompose tens or hundreds.

Model the Skill

Hand out base-ten blocks and write the following problem on the board.

$$28 - 16 =$$

◆ **Say:** *You can use base-ten blocks to model subtraction.* Have students look at the problem and note the subtraction sentence. **Say:** *The beginning number is 28. How do you show 28 with your blocks?* (2 tens, 8 ones) *The number sentence says that we need to subtract 16. How do you show that with the blocks?* (Take away 1 ten and 6 ones.) Guide students to model the subtraction. **Ask:** *What is left?* (12) Observe as students write the difference. Then write another problem on the board.

$$30 - 17 =$$

◆ **Ask:** *What are you going to show with your blocks to model the subtraction in this problem?* (Possible answer: Show 3 tens). *Can you take away 1 ten and 7 ones from 3 tens?* (no) Show students how to trade 1 ten for 10 ones to show 30 in a different way. **Say:** *Now you have some ones to take away. What is left?* (1 ten, 3 ones; 13)

◆ Assign the appropriate practice page(s) to support each student's understanding of the skill.

Assess the Skill

Use the following problems to pre-/post-assess students' understanding of the skill.

$18 - 17 =$	$76 - 27 =$
$34 - 12 =$	$36 - 4 =$
$61 - 31 =$	$47 - 9 =$
$58 - 43 =$	

Name _____

Model the subtraction with blocks. Write the difference.

1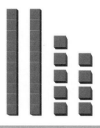

$$29 - 17 = \underline{\hspace{2cm}}$$

2

$$32 - 11 = \underline{\hspace{2cm}}$$

3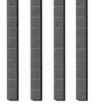

$$40 - 21 = \underline{\hspace{2cm}}$$

4

$$33 - 15 = \underline{\hspace{2cm}}$$

5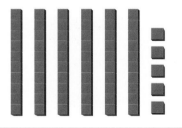

$$65 - 26 = \underline{\hspace{2cm}}$$

6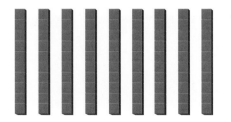

$$90 - 35 = \underline{\hspace{2cm}}$$

 Tell how you regroup blocks to subtract.

Name _____

Find each difference.

1 45 − 34 =

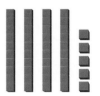

	tens	ones
	4	5
−	3	4

2 66 − 43 =

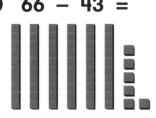

	tens	ones
	6	6
−	4	3

3 59 − 24 =

	tens	ones
−		

4 38 − 18 =

	tens	ones
−		

5 64 − 15 =

	tens	ones
−		

6 74 − 71 =

	tens	ones
−		

7 66 − 20 =

	tens	ones
−		

8 88 − 68 =

	tens	ones
−		

9 30 − 21 =

	tens	ones
−		

10 55 − 17 =

	tens	ones
−		

 Tell how you subtract.

●●○

Name _____

Find each difference.

1 68 – 47 =

2 45 – 23 =

3 39 – 29 =

4 56 – 23 =

5 72 – 41 =

6 67 – 28 =

7 55 – 29 =

8 52 – 23 =

9 70 – 48 =

10 94 – 37 =

11 65 – 29 =

12 82 – 53 =

☆ **Tell how you know when to regroup.**

Name _____

Solve.

1 Mia has 36 hairpins. She uses 14 in her bun. How many pins does she have left?

2 Rachel has 47 cupcakes. She brings 38 to school. How many cupcakes does she have left?

Circle the correct answer for each problem.

3 The Blue Jays had 48 runs last season. The Red Robins had 29. How many more runs did the Blue Jays have?

a) 9 runs

b) 11 runs

c) 19 runs

d) 21 runs

4 Kyra has 75 dollars. She buys some shoes for 39 dollars. How many dollars does she have left?

a) 26 dollars

b) 36 dollars

c) 46 dollars

d) 44 dollars

Unit 14
Subtract Three-Digit Numbers

Standard

Number & Operations in Base Ten

Use place value understanding and properties of operations to add and subtract.

2.NBT.7. Add and subtract within 1,000, using concrete models or drawings and strategies based on place value, properties of operations, and/or the relationship between addition and subtraction; relate the strategy to a written method. Understand that in adding or subtracting three-digit numbers, one adds or subtracts hundreds and hundreds, tens and tens, ones and ones; and sometimes it is necessary to compose or decompose tens or hundreds.

Model the Skill

Hand out base-ten blocks and write the following problem on the board.

$$456 - 214 =$$

◆ **Say:** *Today, we are going to subtract three-digit numbers. Show the first amount with your blocks. What did you show?* (4 hundreds, 5 tens, and 6 ones) *You need to subtract 214. How do you do that?* (Take away 2 hundreds, 1 ten, and 4 ones.) *What do you have left?* (2 hundreds, 4 tens, and 2 ones) Have students write the number on as the difference between the two numbers. You can have students cross out pictured blocks to record the subtraction. Then write another problem on the board.

$$250 - 134 =$$

◆ Have students show 250 with their blocks. **Say:** *We need to subtract 134, but there are not any ones blocks to take away. What do we need to do?* (Possible answer: Trade 1 ten for 10 ones.) Have students make the trade and show them how the blocks still represent the number 250. **Say:** *Now you can subtract. What do you have left?* (116)

◆ Assign students the appropriate practice page(s) to support their understanding of the skill. **Say:** *Use base-ten blocks to model the first number. Take away blocks to show the subtraction, regrouping when needed. Record the subtraction in a place value chart, showing the regrouping.*

Assess the Skill

Use the following problems to pre-/post-assess students' understanding of the skill.

$156 - 32 =$	$429 - 136 =$
$433 - 389 =$	$691 - 350 =$
$576 - 306 =$	$600 - 208 =$
$346 - 147 =$	$715 - 420 =$

Name _____

Model the subtraction with blocks. Write the difference.

❶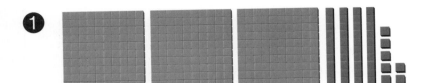

$347 - 217 =$ _____

❷

$324 - 111 =$ _____

❸

$213 - 106 =$ _____

❹

$305 - 210 =$ _____

❺

$438 - 267 =$ _____

❻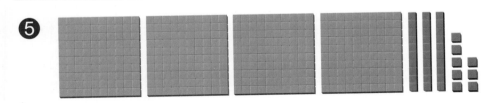

$400 - 272 =$ _____

☆ **Tell how you regroup when there are not enough tens or ones.**

●○○   Unit 14 • Common Core Mathematics Grade 2 • ©2012 Newmark Learning, LL

Find each difference.

1

357 − 142 =

hundreds	tens	ones
3	5	7
− 1	4	2

2

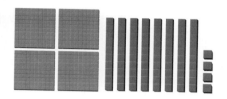

484 − 251 =

hundreds	tens	ones
4	8	4
− 2	5	1

3

989 − 265 =

hundreds	tens	ones
−		

4

812 − 174 =

hundreds	tens	ones
−		

5

529 − 327 =

hundreds	tens	ones
−		

6

654 − 209 =

hundreds	tens	ones
−		

7

700 − 306 =

hundreds	tens	ones
−		

8

578 − 199 =

hundreds	tens	ones
−		

 Tell how you know your answer is reasonable.

Name _____

Find each difference.

1 470 – 247 =   **2** 382 – 149 =   **3** 428 – 143 =

4 319 – 276 =   **5** 967 – 554 =   **6** 791 – 366 =

7 648 – 283 =   **8** 921 – 477 =   **9** 786 – 181 =

10 609 – 158 =   **11** 939 – 275 =   **12** 560 – 182 =

 Tell how you record subtraction with regrouping.

●●●   Unit 14 • Common Core Mathematics Grade 2 • ©2012 Newmark Learning, L

Solve.

1 One year has 365 days. We have school 180 days of the year. How many days of the year do we not have school?

2 The book is 512 pages long. Dave reads 179 pages. How many pages does he have left?

Circle the correct answer for each problem.

3 There are 804 runners in the race. 515 have crossed the finish line. How many runners still have to finish the race?

a) 289 runners

b) 389 runners

c) 269 runners

d) 311 runners

4 The price tag on the oven says 869 dollars. Bill buys it for 70 dollars less. How much does Bill pay for the oven?

a) 169 dollars

b) 799 dollars

c) 899 dollars

d) 859 dollars

Unit 15
Inch, Foot, Yard

Standard

Measurement & Data

Measure and estimate lengths in standard units.

2.MD.1. Measure the length of an object by selecting and using appropriate tools such as rulers, yardsticks, meter sticks, and measuring tapes.

2.MD.2. Measure the length of an object twice, using length units of different lengths for the two measurements; describe how the two measurements relate to the size of the unit chosen.

2.MD.3. Estimate lengths using units of inches, feet, centimeters, and meters.

2.MD.4. Measure to determine how much longer one object is than another, expressing the length difference in terms of a standard length unit.

Model the Skill

Hand out inch rulers.

◆ Display an inch ruler, a yardstick, and a tape measure, and have students identify each object. **Say:** *These are three different tools that can measure the length of an object. Look at your ruler. What units does a ruler use to measure?* (inches and feet) Discuss that a tape measure can also measure in inches and feet while a yardstick can measure in inches, feet, or yards.

◆ **Say:** *When you measure an object, line up the zero mark of the ruler with one edge of the object you are measuring. If there is no zero mark, line up the end of the ruler with the edge of the object. Look to see which inch measurement is closest to the other end of the object. About how long is a typical white sheet of paper?* (about 11 inches)

◆ Assign students the appropriate practice page(s) to support their understanding of the skill. Check that they line up the zero mark of the ruler with the left edge of the object and see which inch the other end of the object is closest to.

Assess the Skill

Use the following problems to pre-/post-assess students' understanding of the skill. Have students use rulers to measure a variety of classroom items.

• Large paper clip—about 2 inches

• Ballpoint pen (with cap)—about 6 inches

• Stapler—about 7 inches

• Book or magazine—answers may vary

Name _____

Measure the length of the object.

1

about _____ inches

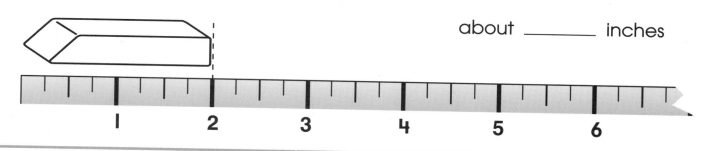

2

about _____ inches

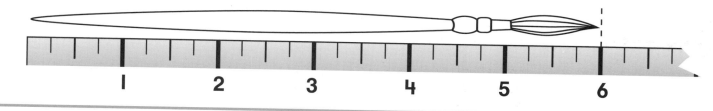

3

about _____ inches

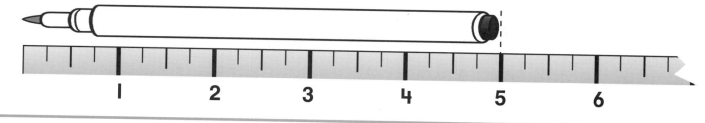

4

about _____ inches

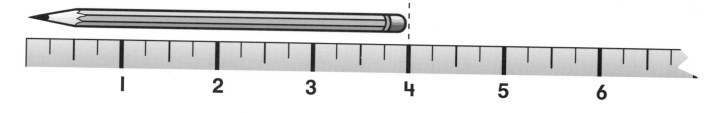

 Tell how you know the length of the object.

Name _____

Measure the length of the object in inches and in feet.

1 a desk

about _____ inches

about _____ feet

2 a door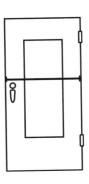

about _____ inches

about _____ feet

3 a board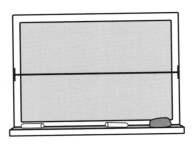

about _____ inches

about _____ feet

4 a classmate

about _____ inches

about _____ feet

5 a table

about _____ inches

about _____ feet

6 a bookshelf

about _____ inches

about _____ feet

 Tell how the two measurements relate to the size of the units.

Estimate the length of the object. Circle the unit. Then measure the object.

1 your pencil

estimate: about _____ inches/feet
measure: about _____ inches/feet

2 a table

estimate: about _____ inches/feet
measure: about _____ inches/feet

3 a crayon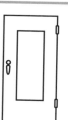

estimate: about _____ inches/feet
measure: about _____ inches/feet

4 a book

estimate: about _____ inches/feet
measure: about _____ inches/feet

5 a door

estimate: about _____ inches/feet
measure: about _____ inches/feet

6 a desk

estimate: about _____ inches/feet
measure: about _____ inches/feet

7 a classmate

estimate: about _____ inches/feet
measure: about _____ inches/feet

8 a window

estimate: about _____ inches/feet
measure: about _____ inches/feet

 Tell how you know your estimate is reasonable.

Name _____

Measure the objects to compare their lengths. Circle the unit.

1 How much longer is this activity page than a crayon?

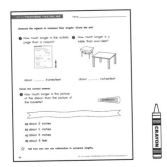

about _____ inches/feet

2 How much longer is a table than your desk?

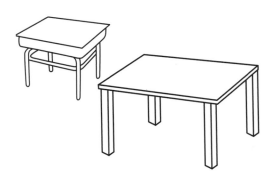

about _____ inches/feet

Circle the correct answer.

3 How much longer is the picture of the ribbon than the picture of the barrette?

a) about 3 inches

b) about 4 inches

c) about 5 inches

d) about 5 feet

☆ **Tell how you can use subtraction to compare lengths.**

Unit 15 • Common Core Mathematics Grade 2 • ©2012 Newmark Learning,

Unit 16
Centimeter, Meter

Standard

Measurement & Data

Measure and estimate lengths in standard units.

2.MD.1. Measure the length of an object by selecting and using appropriate tools such as rulers, yardsticks, meter sticks, and measuring tapes.

2.MD.2. Measure the length of an object twice, using length units of different lengths for the two measurements; describe how the two measurements relate to the size of the unit chosen.

2.MD.3. Estimate lengths using units of inches, feet, centimeters, and meters.

2.MD.4. Measure to determine how much longer one object is than another, expressing the length difference in terms of a standard length unit.

Model the Skill

Hand out centimeter rulers.

◆ Display a centimeter ruler and a meterstick. **Say:** *These are two different tools that can measure the length of an object in metric units. They look similar to the tools used to measure in customary units—inches, feet, and yards—except they measure in metric units. What are some metric units?* (Possible answers: centimeters, meters, and millimeters) Discuss that a meterstick is the length of 100 centimeters.

◆ **Say:** *When you measure an object, you need to make sure that you line up the zero mark of the ruler with one edge of the object you are measuring. Then look to see which centimeter measurement is closest to the other end of the object. About how long is a standard sheet of paper?* (about 28 centimeters)

◆ Assign students the appropriate practice page(s) to support their understanding of the skill. Check that they line up the zero mark of the ruler with the left edge of the object and see which inch the other end of the object is closest to.

Assess the Skill

Use the following problems to pre-/post-assess students' understanding of the skill.

• Large paper clip—about 5 centimeters

• Ballpoint pen (with cap)—about 15 centimeters

• Stapler—about 18 centimeters

• Book or magazine—answers may vary

Name _____

Measure the length of the object.

1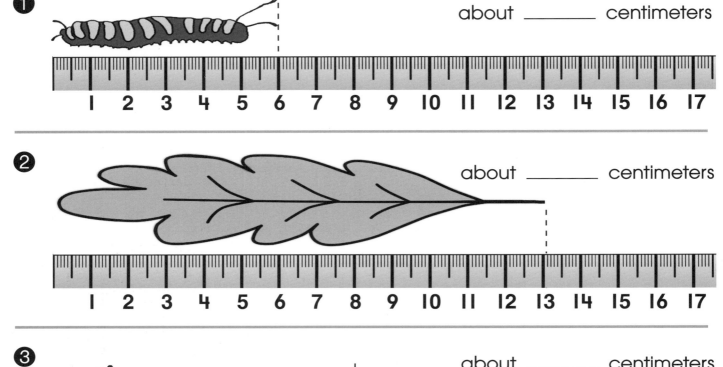

about _____ centimeters

2

about _____ centimeters

3

about _____ centimeters

4

about _____ centimeters

 Tell how you know the length of the object.

Name _____

Measure the object in centimeters and in meters.

1 a table

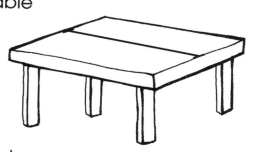

about _____ centimeters

about _____ meters

2 a window

about _____ centimeters

about _____ meters

3 a board

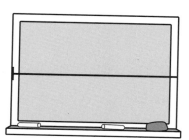

about _____ centimeters

about _____ meters

4 a bookshelf

about _____ centimeters

about _____ meters

5 a classmate

about _____ centimeters

about _____ meters

6 the door

about _____ centimeters

about _____ meters

☆ **Tell how the two measurements relate to the size of the units.**

Name _____

Estimate the length of the object. Circle the unit. Then measure the object.

1 a board eraser

estimate: _____ centimeters/meters
about _____ centimeters/meters

2 a board

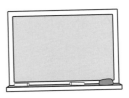

estimate: _____ centimeters/meters
about _____ centimeters/meters

3 your shoe

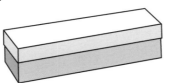

estimate: _____ centimeters/meters
about _____ centimeters/meters

4 a friend

estimate: _____ centimeters/meters
about _____ centimeters/meters

5 a door

estimate: _____ centimeters/meters
about _____ centimeters/meters

6 a book

estimate: _____ centimeters/meters
about _____ centimeters/meters

7 the width of the hallway

estimate: _____ centimeters/meters
about _____ centimeters/meters

8 the length of the room

estimate: _____ centimeters/meters
about _____ centimeters/meters

☆ **Tell how you know your estimate is reasonable.**

Name _____

Measure the objects to compare their lengths. Circle the unit.

1 How much longer is this
activity page than a crayon?

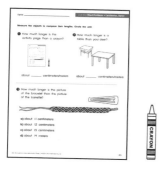

about _____ centimeters/meters

2 How much longer is a
table than your desk?

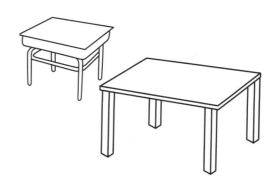

about _____ centimeters/meters

---

3 How much longer is the picture
of the bracelet than the picture
of the barrette?

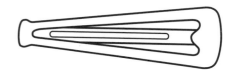

a) about 11 centimeters

b) about 12 centimeters

c) about 13 centimeters

d) about 14 meters

Unit 17
Add and Subtract Lengths

Standard

Measurement & Data

Relate addition and subtraction to length.

2.MD.5. Use addition and subtraction within 100 to solve word problems involving lengths that are given in the same units, e.g., by using drawings (such as drawings of rulers) and equations with a symbol for the unknown number to represent the problem.

2.MD.6. Represent whole numbers as lengths from 0 on a number line diagram with equally spaced points corresponding to the numbers 0, 1, 2, . . . , and represent whole-number sums and differences within 100 on a number line diagram.

Model the Skill

Write the following word problem on the board.

Tom used 9 inches of gold wire and 3 inches of silver wire. How many inches of wire did he use in all?

◆ Hold up a ruler. **Ask:** *How is a number line like a ruler?* (Possible answer: They both show numbers in order that are equally spaced.) *You can use a number line to draw a picture for a word problem.*

◆ Read aloud the word problem. **Say:** *Tom used 9 inches of gold wire. The problem tells us that he also used 3 inches of silver wire. To find out how many inches of wire he used in all, we can add 3 inches to 9 inches. What is 9 plus 3?* (12)

◆ Assign students the appropriate practice page(s) to support their understanding of the skill.

Assess the Skill

Use the following problems to pre-/post-assess students' understanding of the skill.

Susie used 40 centimeters of pink string and 60 centimeters of blue string to make a bracelet.

How many centimeters of string did she use in all?

How many more centimeters of blue string did she use than pink string?

Add and subtract lengths. Use the number line to help you.

1 Lisa used 13 inches of purple string
and 7 inches of pink string. How many
inches of string did Lisa use all together?

_____ inches of string

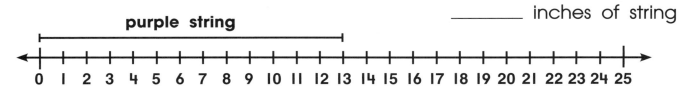

2 Mr. Rocco has a 10-foot piece of garden hose.
He bought a new 5-foot piece of garden hose.
How many feet of garden hose does Mr. Rocco have now?

_____ feet of garden hose

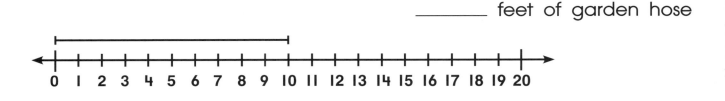

3 Brendon had 20 meters of fishing line.
He gave his dad 6 meters. How many
meters of fishing line did Brendon have left?

_____ meters of line

Tell how you can use a number line to add lengths.

Name _____

Add and subtract lengths. Use a number line to help you.

1 Mark had 10 meters of tape. He used 8 meters of tape.
How many meters of tape does Mark have left?

_____ meters of tape left

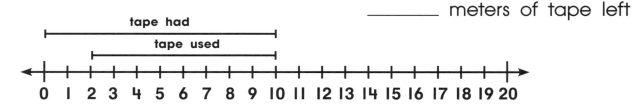

2 Keith bought 18 meters of rope. He had 12 meters of rope.
How many meters of rope does Keith have in all?

_____ meters of rope

3 A ball of string has 18 yards on it. If Rachel uses
5 yards, how many yards of string will be left?

_____ yards of string

4 Teresa has 6 feet of green string and 12 feet of blue string. How
much string does Rachel have in all?

_____ feet of string

5 Kyle walked 20 meters on Tuesday. He walked 17 meters on
Wednesday. How many more meters did he walk on Tuesday
compared to Wednesday?

_____ meters

 Tell how you know the unknown number in the problem.

●●○

Name _____

Add. Complete the number sentence.

1 Donna has 6 yards of black cord and 5 yards of red cord. How many yards of cord does Donna have in all?

6 + 5 = ☐

_____ yards of cord

2 Jim put up a 10-foot ladder. He extended it 9 feet. How many feet does the ladder now reach?

10 + _____ = ☐

_____ feet

3 Scott had an 11-foot piece of carpet. He bought 4 more feet of carpet. How many feet of carpet does he have now?

_____ + _____ = ☐

_____ feet of carpet

4 Barbara put a 9-inch piece of trim on a hat. Then she put on a piece of 8-inch trim. How many inches of trim did Barbara put on the hat?

_____ + _____ = ☐

_____ inches of trim

5 Paul has 14 meters of rope. He cuts off 8 meters and uses it for his boat. How many meters does he have left?

14 – 8 = ☐

_____ meters of rope

6 Before Ava sharpened her pencil, it was 18 centimeters long. After she sharpened the pencil, it was 17 centimeters long. How many centimeters longer was the pencil before she sharpened it?

_____ – _____ = ☐

_____ centimeter(s)

 Tell how you know what number to subtract.

Name _____

Subtract. Complete the number sentence.

1 Amy had a 14-inch piece of sequin trim. She used 5 inches on her costume. How much sequin trim does Amy have left?

2 Tom used a 15-centimeter strip of paper as a bookmark. He cut off 3 centimeters. How long is the bookmark now?

3 The driveway is 19 feet long. The family paved 10 feet. How many feet were left unpaved?

4 Charlie has 9 centimeters of black wire and 7 centimeters of red wire. How much more black wire does Charlie have?

5 Mia's hair was 13 inches long. She had 5 inches cut off. How long is Mia's hair now?

6 Ms. Carrera had 20 meters of ribbon. She used 7 meters to make some wreaths. How many meters of ribbon does she have left?

a) 18 inches

b) 12 inches

c) 6 inches

d) 8 inches

a) 11 meters

b) 12 meters

c) 13 meters

d) 14 meters

Unit 18
Tell Time to the Nearest Five Minutes

Standard

Measurement & Data

Work with time and money.

2.MD.7. Tell and write time from analog and digital clocks to the nearest five minutes, using A.M. and P.M.

Model the Skill

Draw an analog clock that shows 10 o'clock.

◆ **Say:** *You can show the same time in more than one way. What time is shown on this clock?* (10 o'clock) Write **10 o'clock** on the board. **Say:** *There is another way to write **10 o'clock** with only numbers.* Write **10:00. Say:** *The numbers before the colon tell the hour and the numbers after the colon tell the minutes.* Point out that 10:00 says hour 10 and zero minutes.

Draw an analog clock that shows 10:15.

◆ **Say:** *Look at this clock. What time is shown on the digital clock?* (10:15, or a quarter past ten) **Ask:** *How do you know that it is 10:15?*

◆ Assign students the appropriate practice page(s) to support their understanding of the skill.

Assess the Skill

Use the following problems to pre-/post-assess students' understanding of the skill.

Name _____

Circle the two times that are the same. Cross out the time that is different.

1 **12 o'clock**

2:00

2 **7 o'clock**

6:00

3 **9 o'clock**

12:00

4 **9 o'clock**

6:00

5 **5 o'clock**

5:00

6 **5 o'clock**

6:00

☆ **Tell how you know what time is shown.**

●○○

Unit 18 • Common Core Mathematics Grade 2 • ©2012 Newmark Learning, L

Draw lines to match clocks that show the same times.

①    `10:30`

②    `1:00`

③    `1:30`

④    `8:55`

⑤    `12:05`

⑥   `6:45`

⑦   `2:00`

⑧    `4:30`

 Tell how you know when it is five minutes past the hour.

Name _____

Write each time.

1  _____ : _____

2  _____ : _____

3  _____ : _____

4  _____ : _____

5  _____ : _____

6  _____ : _____

7  _____ : _____

8  _____ : _____

9  _____ : _____

10  _____ : _____

11  _____ : _____

12  _____ : _____

 Tell how you tell time to the nearest five minutes.

●●●

Solve.

1 Pipa left for school at 8:25 A.M.
Circle the clock that shows which time she left.

2 David has practice at 3:15 P.M. Circle the clock that shows the time of practice.

Choose the correct answer for each problem.

3 What time is shown below?

a) 7:35 A.M.

b) 7:45 P.M.

c) 7:45 A.M.

d) 9:35 A.M.

4 What time is shown below?

a) 5:40 A.M.

b) 5:40 P.M.

c) 8:25 A.M.

d) 8:25 P.M.

Unit 19
How Much Money?

Standard

Measurement & Data

Work with time and money.

2.MD.8. Solve word problems involving dollar bills, quarters, dimes, nickels, and pennies, using $ and ¢ symbols appropriately. Example: If you have 2 dimes and 3 pennies, how many cents do you have?

Model the Skill

Show examples of several coins including pennies, nickels, dimes, and quarters.

◆ **Ask:** *What types of coins do I have here?* (quarters, dimes, nickels, and pennies) *What is the value of each coin?* (25 cents, 10 cents, 5 cents, and 1 cent) *What is one dime and three pennies?* (13 cents)

◆ Assign students the appropriate practice page(s) to support their understanding of the skill.

Assess the Skill

Use the following problems to pre-/post-assess students' understanding of the skill.

Name _____

Write each amount. Use the ¢ symbol.

1

2

3

4

5

 Tell how you can use skip counting to find amounts of money.

Name _____

Write each amount. Use the ¢ symbol.

1

_____

2

_____

3

_____

4

_____

5

_____

6

_____

7

_____

8

_____

 Tell how you count three different kinds of coins.

Unit 19 • Common Core Mathematics Grade 2 • ©2012 Newmark Learning, L

Match each money to each tag.

1    $3.01

2   $1.26

3   $1.60

4   $0.65

5    $1.41

6   $0.40

7   $2.06

8    $0.41

☆ **Tell how you count on to find the total amount of money.**

Name _____

Solve.

1 You have two dimes, two nickels, and two pennies. What amount of money do you have?

2 Molly has two quarters, one dime, and one nickel. How much money does Molly have?

Choose the correct answer for each problem.

3 You have two one-dollar bills, a nickel, and two pennies. What amount of money do you have?

a) $1.52

b) $2.52

c) $1.07

d) $2.07

4 You have one one-dollar bill, three dimes, and five pennies. What amount of money do you have?

a) $1.35

b) $1.45

c) $1.08

d) $1.53

Unit 19 • Common Core Mathematics Grade 2 • ©2012 Newmark Learning, L

Unit 20
Make a Line Plot

Standard

Measurement & Data

Represent and interpret data.

2.MD.9. Generate measurement data by measuring lengths of several objects to the nearest whole unit, or by making repeated measurements of the same object. Show the measurements by making a line plot, where the horizontal scale is marked off in whole-number units.

Model the Skill

Draw this tally chart and corresponding line plot on the board.

Seedlings in Ms. Goya's Class	
Height in Centimeters	Number
4	卌
5	‖‖
6	卌 ‖‖

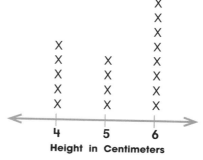

Height in Centimeters

◆ **Say:** *Data can be shown in charts and graphs. A line plot uses Xs to show data on a number line. What does this line plot show?* (the heights of the seedlings in Ms. Goya's class) *What are the different heights of the students?* (4, 5, and 6 centimeters) Have students point to the scale that shows the heights and see the labels.

◆ **Say:** *The Xs show how many seedlings are each height. How can you find the most common height of the seedlings in the class?* (Look for the height that has the most Xs.) **Ask:** *Which height has the most Xs?* (6 centimeters) *Most of the seedlings are 6 centimeters tall.*

◆ **Ask:** *How do you find the number of seedlings that are 4 centimeters tall?* (Look for the 4 on the scale and count the number of Xs above it.) *What number did you count?* (5)

◆ Assign students the appropriate activity page(s) to support their understanding of the skill.

Assess the Skill

Ask students to measure each other's height and create a tally chart. Then ask them to use the height data to create a line plot.

Name _____

Solve each problem. Use the data in the line plot.

The line plot shows the heights of the students in Mrs. Smith's class.

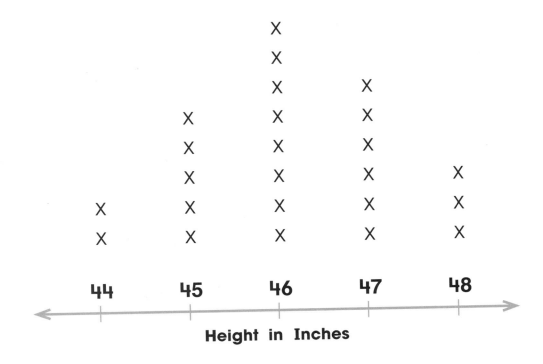

1 What is the most common height of the students in the class? _____

_____

2 How many students are 45 inches tall? _____

_____

3 How many students are 48 inches tall? _____

_____

4 Which height shows 6 students? _____

☆ **Tell how you read a line plot.**

Name _____

Use the data to complete the line plots.

1

Seedlings in Mr. Falber's Class	
Height in Centimeters	Number
4	IIII
5	IIII I
6	IIII III
7	II

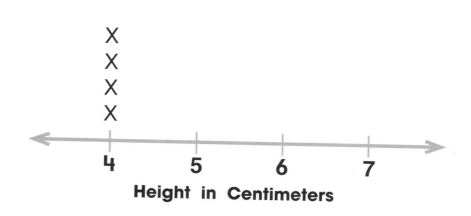

Height in Centimeters

What is the least common height of the seedlings? _____

2

Height of Sunflowers		
4 feet	4 feet	5 feet
5 feet	3 feet	6 feet
3 feet	4 feet	6 feet
3 feet	5 feet	5 feet
3 feet	6 feet	5 feet
5 feet	3 feet	6 feet

```
←————|————|————|————|————→
     3    4    5    6
```

Height in Feet

How many sunflowers are more than 4 feet tall? _____

☆ **Tell how you completed a line plot.**

Name _____

Use the data to complete the line plots.

1

Height of Bean Plants		
10 cm	9 cm	7 cm
8 cm	9 cm	9 cm
8 cm	9 cm	7 cm
9 cm	10 cm	8 cm
7 cm	8 cm	8 cm
10 cm	10 cm	9 cm

Height in Centimeters
_____

What is the most common height for the sunflowers? _____

2

Height of Cornstalks		
8 feet	7 feet,	5 feet
8 feet	8 feet	6 feet
8 feet	7 feet	6 feet
8 feet	7 feet	8 feet
6 feet	5 feet	6 feet
8 feet	6 feet	

_____

What is the least common height for the cornstalks? _____

3

Height of Tomato Plants		
34 in	35 in	33 in
33 in	34 in	32 in
33 in	34 in	32 in
34 in	33 in	34 in
32 in	33 in	35 in
34 in	33 in	

_____

How many tomato plants are less than 34 inches tall? _____

 Tell how you know that you have plotted all the data.

Name _____

Use the data to complete the line plot.

1 Make a line plot to show the lengths of the necklaces that Kylee made.

Length of Necklaces	
15 inches	17 inches
16 inches	18 inches
18 inches	16 inches
15 inches	18 inches
18 inches	15 inches

_____

2 What is the most common length for the bracelets? _____

3 What is the least common length for the bracelets? _____

4 How many bracelets are more than 16 inches long?

_____

a) 2
b) 3
c) 4
d) 5

5 How many bracelets did Kylee make in all? _____

a) 8
b) 9
c) 10
d) 12

 Tell how you know what to label the scale.

Unit 21
Make a Graph

Standard

Measurement & Data

Represent and interpret data.

2.MD.10. Draw a picture graph and a bar graph (with single-unit scale) to represent a data set with up to four categories. Solve simple put- together, take-apart, and compare problems using information presented in a bar graph.

Model the Skill

Draw the following bar graph.

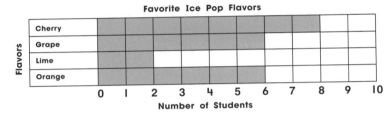

- ◆ **Say:** *Graphs can be used to show data. What does the bar graph show?* (Favorite Ice Pop Flavors) Have students look at the labels. **Say:** *One axis tells us the different flavors and the other shows a scale of numbers. What are the flavors?* (Cherry, Grape, Lime, and Orange) *To solve the first problem, find out how many like cherry. Slide your finger on the bar labeled* **Cherry** *and then down to the number scale. The bar is at what number?* (8) Repeat the process for grape. (6) **Say:** *To find how many students like cherry and grape, add the numbers.* Write the number sentence. **Ask:** *How many students like cherry and grape ice pops?* (14)

- ◆ **Ask:** *What number are you going to subtract from 22 to find how many students like another flavor better than lime?* (the number of students that like lime—2) Have students complete the number sentence to find the difference. (22 – 2 = 20)

- ◆ Now find the number of students that like orange and the number that like lime. Write a number sentence to compare the two amounts. (Answers: 6; 2; possible equation: 6 – 2 = 4)

- ◆ Assign students the appropriate activity page(s) to support their understanding of the skill.

Assess the Skill

Use the following problems to pre-/post-assess students' understanding of the skill.

Ask students to make a survey of the class shoe sizes and make a graph that shows shoe size data.

Name _____

Use the graph to solve each problem.

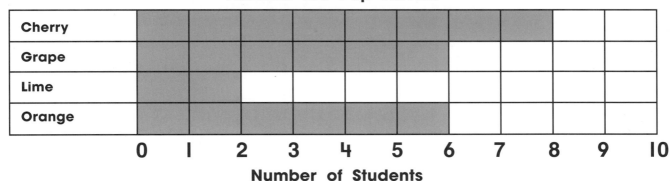

Favorite Ice Pop Flavors

Flavors											
Cherry											
Grape											
Lime											
Orange											
	0	1	2	3	4	5	6	7	8	9	10

Number of Students

❶ How many students like cherry and grape ice pops?

_____ students like cherry. _____ students like grape.

_____ + _____ = _____ students like cherry and grape.

❷ How many students like another flavor other than lime?

_____ + _____ + _____ = _____

_____ students like another flavor better than lime.

❸ How many more students like orange than like lime?

_____ students like orange. _____ students like lime.

_____ ◯ _____ = _____

_____ more students like orange than lime.

❹ How many more students like cherry than like lime?

_____ students like cherry. _____ students like lime.

_____ ◯ _____ = _____

_____ more students like cherry than lime.

 Write how many students are in the survey. Tell how you found the answer.

●○○ 107

Name _____

Use the data to complete the bar graph. Then use the graph to solve the problems.

Favorite Subjects	
Subject	**Number**
Math	⊔⊓⊤ ⊔⊓⊤
Reading	⊔⊓⊤ |||
Science	||||
Social Studies	|||

Favorite Subject

Subject											
Math											
Reading											
Science											
Social Studies											

0 1 2 3 4 5 6 7 8 9 10

Number of Students

1 A total of 25 students are shown on the graph.
How many students like another subject better than math?

_____ students like math.   25 − _____ = _____

_____ students like another subject better than math.

---

2 How many students like reading and science?

_____ students like reading.   _____ students like science.

_____ ◯ _____ = _____

_____ students like reading and science.

---

3 How many more students like math than like reading?

_____ ◯ _____ = _____

_____ more students like math than reading.

---

4 How many more students like math than like social studies?

_____ ◯ _____ = _____

_____ more students like math than social studies.

 Tell how the graph helps you see the data better than a list or chart.

●●◯

Use the data to complete the bar graphs. Then use the graphs to answer each question.

1

Favorite Sports	
Sport	**Number**
Soccer	JHH I
Baseball	II
Basketball	IIII
Football	JHH II

Favorite Sports

Soccer										
Baseball										
Basketball										
Football										

0 1 2 3 4 5 6 7 8 9 10

Number of _____

2 How many students like soccer better than basketball? _____

3 How many students like baseball and football? _____

4 How many students like a sport other than baseball? _____

5

Pets We Have	
Pet	**Number**
Dog	JHH JHH
Cat	JHH III
Fish	II

_____

Dog										
Cat										
Fish										

0 1 2 3 4 5 6 7 8 9 10

_____

6 How many more students like dogs more than cats? _____

7 How many more students like dogs more than fish? _____

8 How many more students prefer dogs to cats and fish? _____

☆ **Tell how you know what to label each axis of the graph.**

Name _____

Use the data to complete the graphs.

1

Favorite Kinds of Games	
Games	**Number**
Card	ЖЖ I
Board	III
Outdoor	ЖЖ III

_____

0   1   2   3   4   5   6   7   8

_____

2

Favorite Colors	
Color	**Number**
Red	IIII
Blue	ЖЖ II
Yellow	III
Green	ЖЖ I

Favorite Colors

Red	☺ ☺ ☺ ☺
Blue	
Yellow	
Green	

Key ☺ = 1 student

3

Favorite Snacks	
Snack	**Number**
Pretzels	5
Yogurt	3
Fruit	8

_____

_____	
_____	
_____	

Key _____ = _____

Unit 22
Identify Shapes

Standard

Geometry

Reason with shapes and their attributes.

2.G.1. Recognize and draw shapes having specified attributes, such as a given number of angles or a given number of equal faces. Identify triangles, quadrilaterals, pentagons, hexagons, and cubes.

Model the Skill

Hand out tangrams and cubes.

◆ Display models of triangles, quadrilaterals, pentagons, hexagons, and cubes. **Ask:** *Which one of these shapes is most different from the other shapes?* (a cube) Hold up a cube and explain that it is a three-dimensional shape while the other shapes are two-dimensional, or flat, shapes. **Ask:** *How many flat sides, or faces, does a cube have?* Have two volunteers work together to determine the number of faces. (6) **Ask:** *What shape are the faces of a cube?* (square)

◆ Hold up a triangle. **Say:** *Use a blue crayon to trace the sides of the triangle. How many sides did you trace?* (3) Allow tactile learners to touch the triangle models. **Say:** *To count the angles, use a red crayon to make a mark on each angle. How many angles are there?* (3) Explain to students that *tri-* means "three," and a triangle is a shape with three angles.

◆ Have students trace and mark the sides of a quadrilateral. **Say:** *Think about other words that you know that begin with* quad. Quad *means "four." A quadrilateral has four sides and four angles.* Point out various quadrilaterals including a square, a rectangle, a rhombus, a trapezoid, and other parallelograms.

◆ Assign students the appropriate practice page(s) to support their understanding of the skill. Guide students to link the names of each shape to the number of sides and angles of each shape.

Assess the Skill

Have students pick up handfuls of tangram shapes and ask them to trace and name each one.

Name _____

List the number of sides for each shape. Then list the number of angles.

1 **triangle**

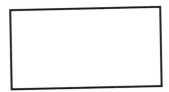

sides _____

angles _____

2 **quadrilateral**

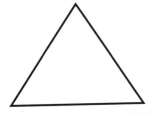

sides _____

angles _____

3 **pentagon**

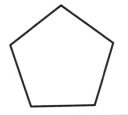

sides _____

angles _____

4 **hexagon**

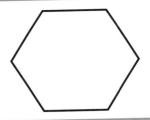

sides _____

angles _____

5 **square**

sides _____

angles _____

6 **octagon**

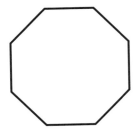

sides _____

angles _____

☆ **Tell how the number of sides matches the shape name.**

●○○

Name _____

Match each shape to its name.

1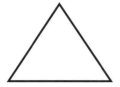

parallelogram

2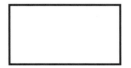

triangle

3

rectangle

4

cube

5

hexagon

6

pentagon

 Circle all of the quadrilaterals.

●●○   113

Name _____

Label each shape.

1  _____

2  _____

3 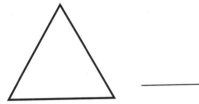 _____

4 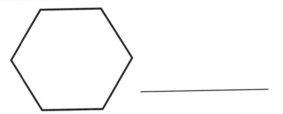 _____

5 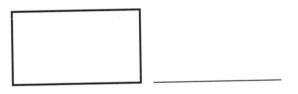 _____

6 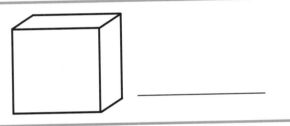 _____

7 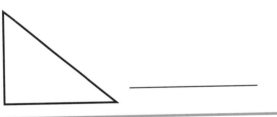 _____

8 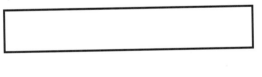 _____

9 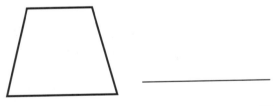 _____

10 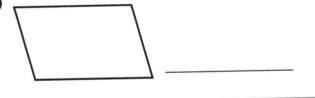 _____

11 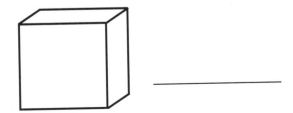 _____

12 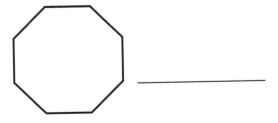 _____

 Tell how you know which shapes are quadrilaterals.

Draw each shape described. Write its name.

1 I have four equal sides and four equal angles. What am I?

2 I have six sides and six angles. What am I?

_____

_____

3 I have four sides and four angles. What am I?

4 I have six faces and six angles. What am I?

_____

_____

5 I have three sides and three angles. What am I?

6 I have five sides and five angles. What am I?

_____

_____

 **Tell how you know which shape is described.**

Unit 23
Parts of Shapes

Standard

Geometry

Reason with shapes and their attributes.

2.G.2. Partition a rectangle into rows and columns of same-size squares and count to find the total number of them.

2.G.3. Partition circles and rectangles into two, three, or four equal shares, describe the shares using the words *halves, thirds, half of, a third of,* etc., and describe the whole as two halves, three-thirds, four-fourths. Recognize that equal shares of identical wholes need not have the same shape.

Model the Skill

Hand out a rectangular sheet of paper.

◆ **Say:** *Imagine that you and a friend both want to draw, but you have only one sheet of paper. What can you do so that you each have an equal share of the paper?* (Answers will vary. Possible answer: Fold the paper in half and cut along the fold.) Guide students to fold one sheet of paper in half, matching corners. **Say:** *Each share of the paper is the same size. The shares are equal. How is this paper divided?* (into halves) Have students draw a line down the fold to show two equal shares. **Say:** *Each half of the rectangle is one-half. Two halves are the same as one whole.*

◆ **Say:** *We can fold the paper again to make four equal shares.* Guide students to match corners, fold, and then open the paper to see the four sections. **Say:** *The paper shows fourths. Four-fourths are the same as one whole.* Guide students to draw lines on the new fold to show four equal shares.

◆ Assign students the appropriate practice page(s) to support their understanding of the skill. If necessary, have students fold another sheet of paper to see the equal shares for halves, fourths, and thirds.

Assess the Skill

Use the following problems to pre-/post-assess students' understanding of the skill.

Ask students to name the number of equal shares in each shape.

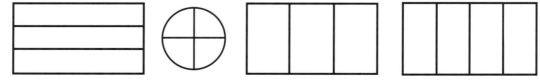

Draw lines to show equal shares.

1 Show two equal shares.

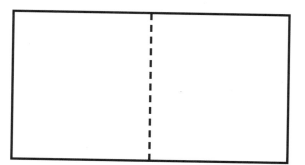

2 Show two equal shares.

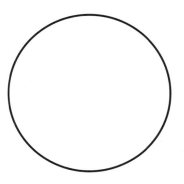

3 Show four equal shares.

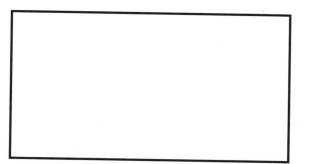

4 Show three equal shares.

5 Show four equal shares.

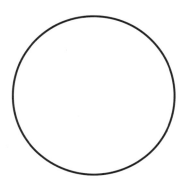

6 Show four equal shares.

☆ **Circle the shapes that show halves. Underline the shape that shows thirds.**

Name _____

1 Show four equal shares.

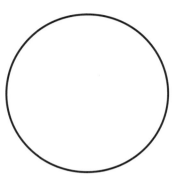

2 Show two equal shares.

3 Show fourths.

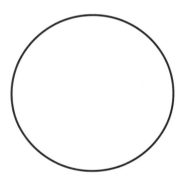

4 Show thirds.

Circle the shapes that shows halves.

5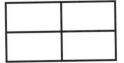

Circle the shapes that shows fourths, or quarters.

6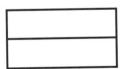

☆ **Tell how equal shares of the same shape can be different shapes.**

Name _____

Circle the shapes that show equal shares. Then label what they show.

❶ 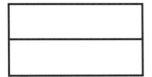 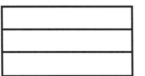 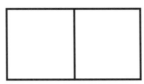 _____

❷ 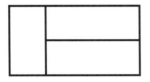 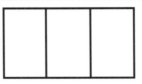 _____

❸ 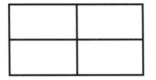 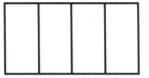 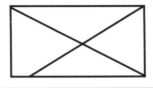 _____

❹ 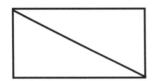 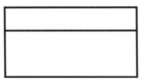 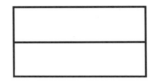 _____

❺   _____

❻   _____

❼   _____

❽   _____

 Tell how you know which description matches each shape.

Name _____

Color to show the shares.

1 one-half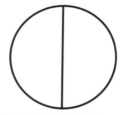

2 one-fourth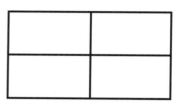

3 one-third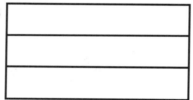

4 two-fourths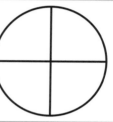

5 one-half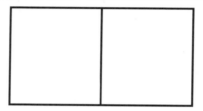

6 one-fourth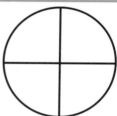

7 two-thirds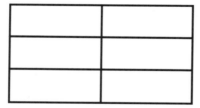

8 two-fourths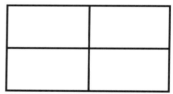

9 one-half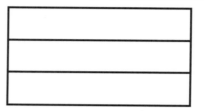

10 three-fourths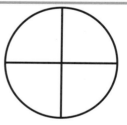

11 one-third

12 two-thirds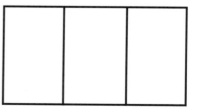

☆ **Tell how two-fourths is like one-half.**

Complete the box for each fact family.

1.
```
        12
     5       7
   ___ + ___ = ___
   ___ + ___ = ___
   ___ - ___ = ___
   ___ - ___ = ___
```

2.
```
        15
     7       8
   ___ + ___ = ___
   ___ + ___ = ___
   ___ - ___ = ___
   ___ - ___ = ___
```

3.
```
        13
     6       7
   ___ + ___ = ___
   ___ + ___ = ___
   ___ - ___ = ___
   ___ - ___ = ___
```

4.
```
        14
     5       9
   ___ + ___ = ___
   ___ + ___ = ___
   ___ - ___ = ___
   ___ - ___ = ___
```

5.
```
        13
     4       9
   ___ + ___ = ___
   ___ + ___ = ___
   ___ - ___ = ___
   ___ - ___ = ___
```

6.
```
        16
     7       9
   ___ + ___ = ___
   ___ + ___ = ___
   ___ - ___ = ___
   ___ - ___ = ___
```

7.
```
        12
     3       9
   ___ + ___ = ___
   ___ + ___ = ___
   ___ - ___ = ___
   ___ - ___ = ___
```

8.
```
        11
     4       7
   ___ + ___ = ___
   ___ + ___ = ___
   ___ - ___ = ___
   ___ - ___ = ___
```

9.
```
        15
     6       9
   ___ + ___ = ___
   ___ + ___ = ___
   ___ - ___ = ___
   ___ - ___ = ___
```

10.
```
        17
     8       9
   ___ + ___ = ___
   ___ + ___ = ___
   ___ - ___ = ___
   ___ - ___ = ___
```

11.
```
        11
     5       6
   ___ + ___ = ___
   ___ + ___ = ___
   ___ - ___ = ___
   ___ - ___ = ___
```

12.
```
        12
     8       4
   ___ + ___ = ___
   ___ + ___ = ___
   ___ - ___ = ___
   ___ - ___ = ___
```

Name _____

Solve.

1   40 − 8 = _____   **2**   15 − 7 = _____

3   32 − 4 = _____   **4**   24 − 9 = _____

5   17 − 9 = _____   **6**   14 − 8 = _____

7   26 − 8 = _____   **8**   42 − 5 = _____

9   53 − 6 = _____   **10** 67 − 3 = _____

11 83 − 9 = _____   **12** 76 − 7 = _____

13 32 − 9 = _____   **14** 30 − 7 = _____

Common Core Mathematics Grade 2 • ©2012 Newmark Learning, L

Solve.

1 68 – 47 = _____

2 45 – 23 = _____

3 39 – 29 = _____

4 56 – 23 = _____

5 72 – 41 = _____

6 67 – 28 = _____

7 55 – 29 = _____

8 52 – 23 = _____

9 70 – 48 = _____

10 94 – 37 = _____

11 65 – 29 = _____

12 82 – 53 = _____

Name _____

Write if the amount is odd or even.

1

2

3

4

5

6

7

8

9

10

11

12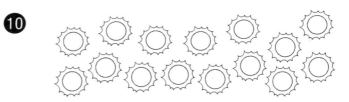

Common Core Mathematics Grade 2 • ©2012 Newmark Learning,

Solve.

1 12 + 48 = _____

2 55 + 16 = _____

3 34 + 22 + 18 = _____

4 41 + 13 + 27 = _____

5 54 + 16 + 24 = _____

6 42 + 13 + 45 = _____

7 21 + 57 + 19 = _____

8 26 + 43 + 11 = _____

9 43 + 27 + 18 = _____

10 37 + 18 + 25 = _____

Name _____

Solve.

1   82 + 17 = _____

2   5 + 67 = _____

3   13 + 9 + 68 = _____

4   22 + 19 + 57 = _____

5   64 + 23 + 13 = _____

6   8 + 13 + 54 = _____

7   8 + 77 + 9 = _____

8   6 + 77 + 20 = _____

9   28 + 38 + 32 = _____

10   58 + 19 + 19 = _____

Common Core Mathematics Grade 2 • ©2012 Newmark Learning,

Name _____

Compare. Write <, >, =.

1 647 ◯ 674

2 639 ◯ 936

3 718 ◯ 781

4 216 ◯ 207

5 354 ◯ 354

6 507 ◯ 520

7 654 ◯ 954

8 309 ◯ 903

9 128 ◯ 134

10 775 ◯ 689

11 422 ◯ 224

12 709 ◯ 699

Name _____

Compare. Write <, >, =.

1 445 ◯ 454

2 322 ◯ 292

3 809 ◯ 908

4 520 ◯ 250

5 763 ◯ 678

6 369 ◯ 369

7 609 ◯ 690

8 348 ◯ 384

9 519 ◯ 489

10 610 ◯ 601

11 478 ◯ 487

12 899 ◯ 901

Common Core Mathematics Grade 2 • ©2012 Newmark Learning,

Solve.

1   652 + 132

2   569 + 329

3   473 + 164

4   148 + 302

5   321 + 149

6   467 + 203

7   730 + 206

8   657 + 308

9   94 + 449

10   272 + 345

11   711 + 367

12   819 + 52

13   368 + 82

14   229 + 417

Name _____

Solve.

❶ 243 + 607

❷ 417 + 347

❸ 280 + 531

❹ 776 − 728

❺ 457 + 502

❻ 899 − 431

❼ 647 − 59

❽ 186 + 195

❾ 316 + 287

❿ 958 − 643

⓫ 513 − 208

⓬ 27 + 456

⓭ 712 + 128

⓮ 319 + 595

Common Core Mathematics Grade 2 • ©2012 Newmark Learning,

Solve.

1 552 − 132

2 669 − 329

3 573 − 164

4 648 − 302

5 321 − 149

6 467 − 203

7 730 − 206

8 657 − 308

9 994 − 49

10 672 − 345

11 711 − 367

12 819 − 52

13 368 − 82

14 527 − 418

Name _____

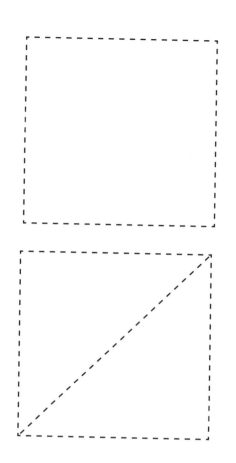

Name _____

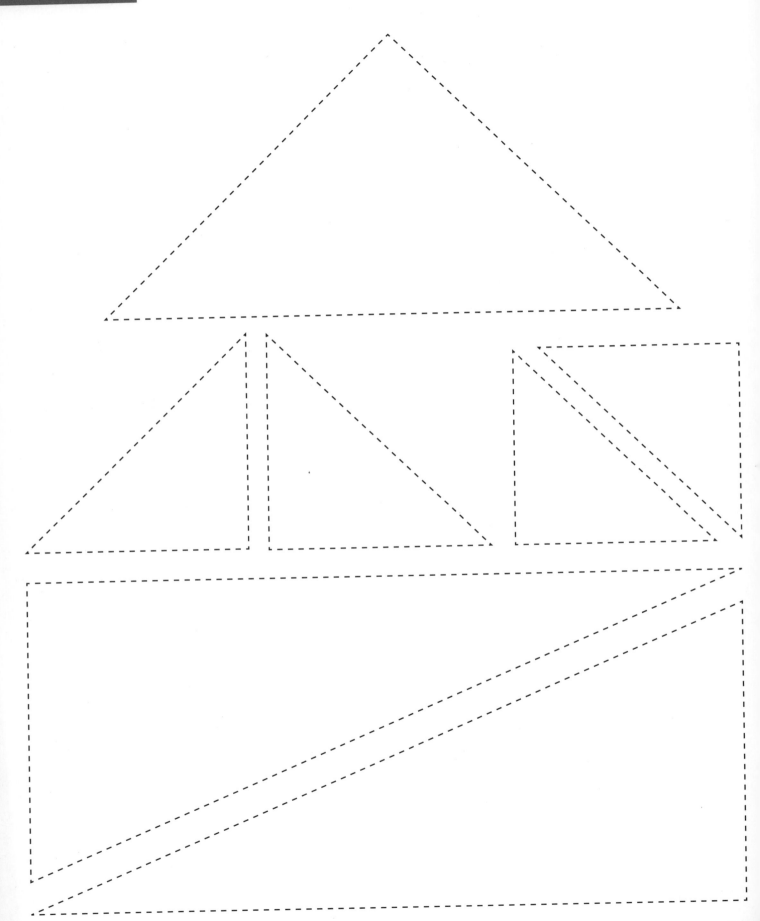

Common Core Mathematics Grade 2 • ©2012 Newmark Learning, LL

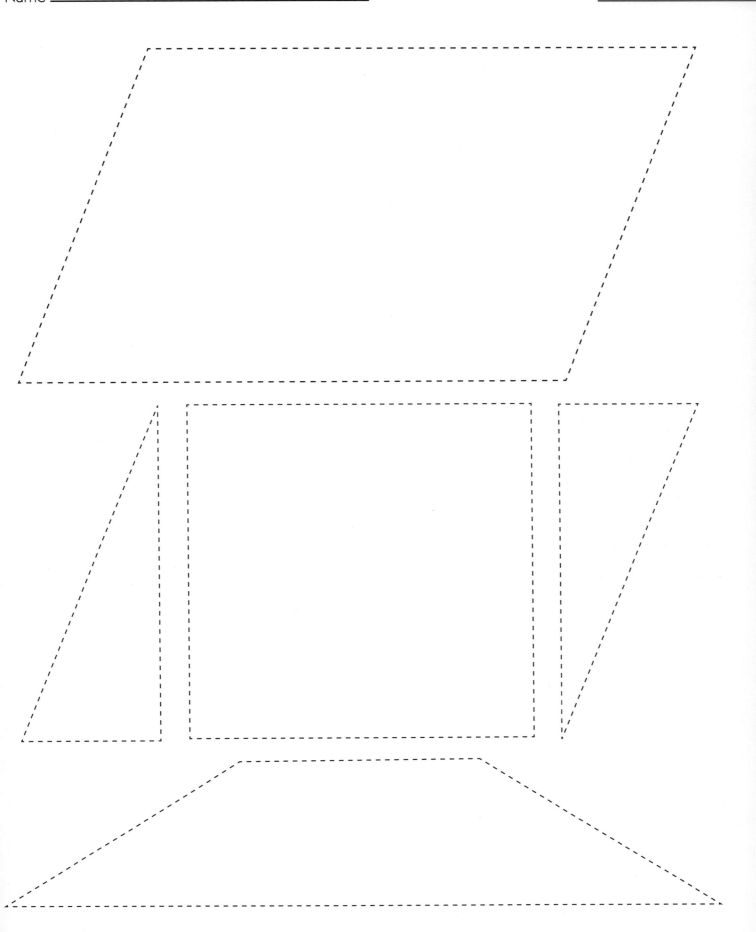

Name _____

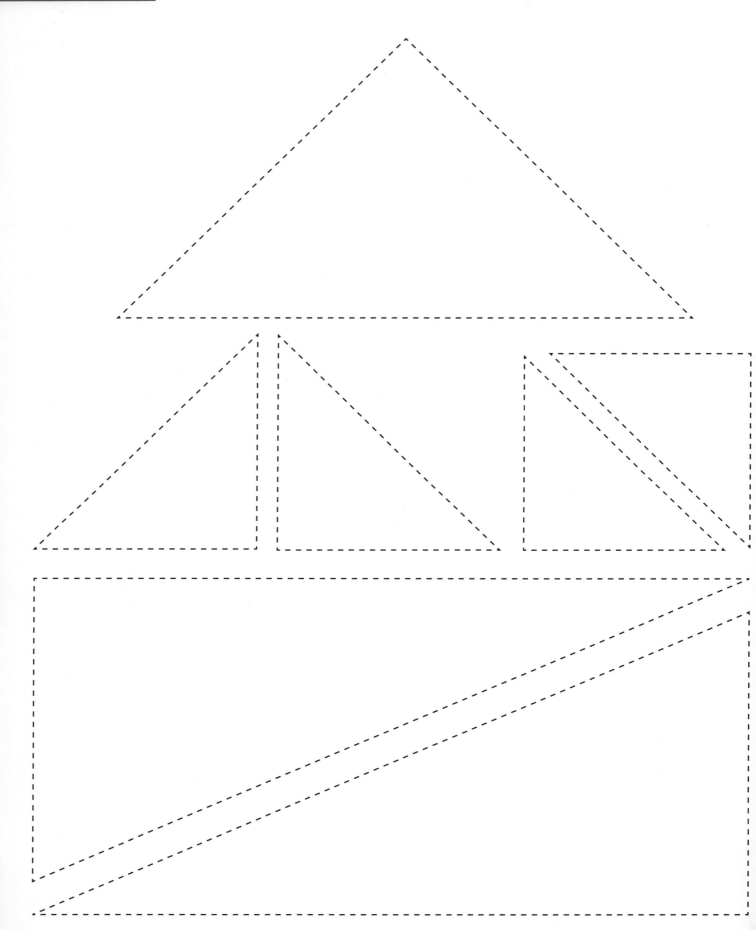

Common Core Mathematics Grade 2 • ©2012 Newmark Learning, L

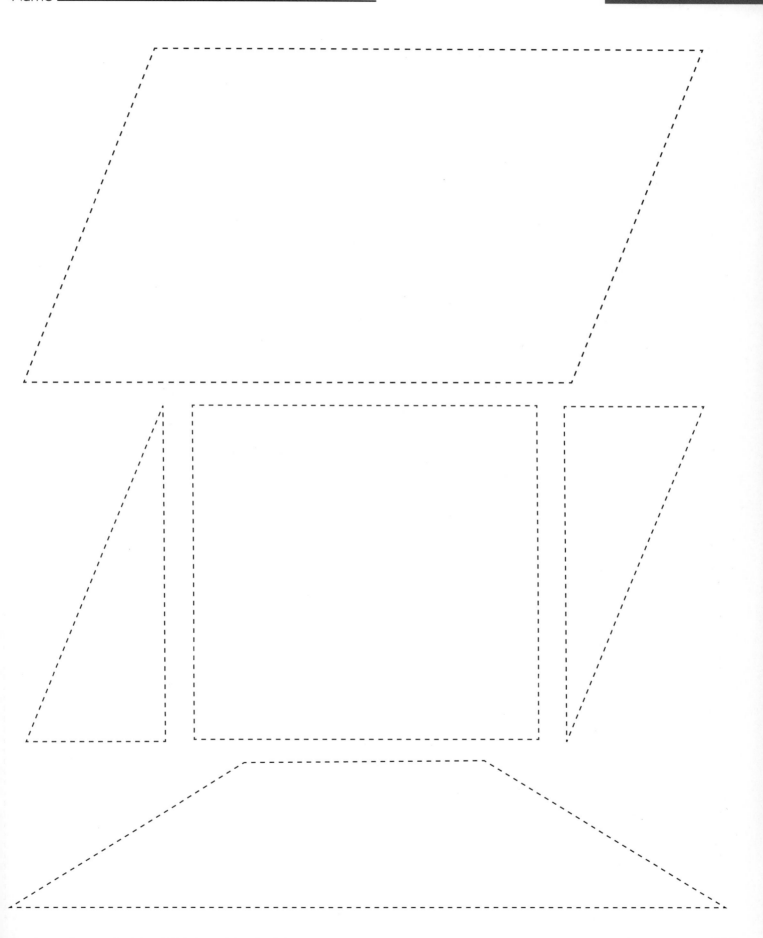

Name _____

Answer Key • Units 1–3

Unit 1 (p. 7)

•

1. $8 - 3 = 5$
2. $10 - 7 = 3$
3. $9 - 8 = 1$
4. $9 - 6 = 3$
5. $6 + 4 = 10$, $10 - 6 = 4$, or $10 - 4 = 6$
6. $7 + 2 = 9$, $9 - 7 = 2$, or $9 - 2 = 7$

Unit 1 (p. 8)

••

1. $2 + 8 = 10$, $10 - 8 = 2$, $10 - 2 = 8$
2. $4 + 8 = 12$, $12 - 8 = 4$, $12 - 4 = 8$

3. $8 + 7 = 15$, $15 - 7 = 8$, $15 - 8 = 7$
4. $6 + 5 = 11$, $11 - 6 = 5$, $11 - 5 = 6$
5. $9 + 3 = 12$, $12 - 9 = 3$, $12 - 3 = 9$
6. $7 + 3 = 10$, $10 - 7 = 3$, $10 - 3 = 7$
7. $10 + 5 = 15$, $15 - 5 = 10$, $15 - 10 = 5$
8. $7 + 5 = 12$, $12 - 7 = 5$, $12 - 5 = 7$
9. $6 + 8 = 14$, $14 - 8 = 6$, $14 - 6 = 8$

Unit 1 (p. 9)

•••

1. $5 + 7 = 12$, $7 + 5 = 12$, $12 - 7 = 5$, $12 - 5 = 7$
2. $8 + 7 = 15$, $7 + 8 = 15$, $15 - 8 = 7$, $15 - 7 = 8$
3. $7 + 6 = 13$, $6 + 7 = 13$, $13 - 7 = 6$, $13 - 6 = 7$
4. $5 + 9 = 14$, $9 + 5 = 14$, $14 - 9 = 5$, $14 - 5 = 9$
5. $4 + 9 = 13$, $9 + 4 = 13$, $13 - 9 = 4$, $13 - 4 = 9$
6. $7 + 9 = 16$, $9 + 7 = 16$, $16 - 9 = 7$, $16 - 7 = 9$
7. $3 + 9 = 12$, $9 + 3 = 12$, $12 - 9 = 3$, $12 - 3 = 9$
8. $4 + 7 = 11$, $7 + 4 = 11$, $11 - 7 = 4$, $11 - 4 = 7$

9. $6 + 9 = 15$, $9 + 6 = 15$, $15 - 6 = 9$, $15 - 9 = 6$
10. $8 + 9 = 17$, $9 + 8 = 17$, $17 - 9 = 8$, $17 - 8 = 9$
11. $5 + 6 = 11$, $6 + 5 = 11$, $11 - 6 = 5$, $11 - 5 = 6$
12. $4 + 8 = 12$, $8 + 4 = 12$, $12 - 8 = 4$, $12 - 4 = 8$

Unit 1 (p. 10)

Word Problems

1. $5 + 7 = 12$, $7 + 5 = 12$ $12 - 7 = 5$
2. $8 + 6 = 14$, $14 - 8 = 6$, $14 - 6 = 8$
3. a
4. b

Unit 2 (p. 12)

•

1. $5 + 4 = 9$
2. $9 - 8 = 1$
3. $7 - 4 = 3$
4. $8 + 4 = 12$, $12 - 2 = 10$

Unit 2 (p. 13)

••

1. $12 - 8 = 4$
2. $9 + 7 = 6$
3. $16 - 5 = 11$
4. $13 - 7 = 6$
5. $11 + 8 = 19$, $19 - 2 = 17$
6. $14 + 3 = 17$, $17 - 4 = 13$

Unit 2 (p. 14)

•••

1. $14 - 6 = 8$
2. $8 + 5 = 13$
3. $9 + 6 = 15$
4. $10 - 8 = 2$
5. $6 + 8 = 14$, $14 - 3 = 11$
6. $16 - 3 = 13$, $13 - 2 = 11$
7. $7 + 2 = 9$, $9 - 4 = 5$
8. $20 - 6 = 14$, $14 - 7 = 7$

Unit 2 (p. 15)

Word Problems

1. $15 - 7 = 8$
2. $6 + 5 = 11$, $11 - 3 = 8$
3. $16 + 3 = 19$
4. $5 + 7 = 12$, $12 - 8 = 4$
5. a
6. c

Unit 3 (p. 17)

•

1. odd   2. even
3. even   4. odd
5. even   6. even
7. even   8. even

Unit 3 (p. 18)

••

1. odd   2. even
3. even   4. odd
5. even   6. even
7. even   8. even
9. even   10. even

Unit 3 (p. 19)

•••

1. even   2. even
3. odd   4. odd
5. even   6. even
7. even   8. even
9. even   10. even
11. even   12. even

Unit 3 (p. 20)

Word Problems

1. 16, even
2. 18, even
3. 19, odd
4. 15, odd
5. b
6. c

Common Core Mathematics Grade 2 • ©2012 Newmark Learning, LI

Answer Key • Units 4–7

Unit 4 (p. 22)
•
1. 4   2. 6
3. 8   4. 10
5. 12   6. 12

Unit 4 (p. 23)
••
1. 6   2. 6
3. 9   4. 15
5. 16   6. 12
7. 12   8. 25

Unit 4 (p. 24)
•••
1. 3 + 3 = 6
2. 2 + 2 + 2 = 6
3. 2 + 2 + 2 + 2 = 8
4. 3 + 3 + 3 = 9
5. 4 + 4 + 4 = 12
6. 3 + 3 + 3 + 3 = 12
7. 4 + 4 + 4 + 4 = 16
8. 3 + 3 + 3 + 3 + 3 = 15
9. 5 + 5 + 5 = 15
10. 4 + 4 + 4 + 4 + 4 = 20

Unit 4 (p. 25)
Word Problems
1. 15 dollars
2. 24 peppers
3. 36 plants
4. 50 raisins
5. c
6. c

Unit 5 (p. 27–29)
Check students' work.

Unit 5 (p. 30)
Word Problems
1. 3 hundreds, 2 tens, 8 ones
2. 3 hundreds, 4 tens, 0 ones
3. 1 hundred, 6 tens, 1 one
4. 5 hundreds, 0 tens, 9 ones
5. d
6. c

Unit 6 (p. 32)
•
1. 300 + 40 + 0, three hundred forty
2. 712, 700 + 10 + 2, seven hundred twelve
3. 300, 400, 500, 600, 700, 800, 900, 1,000
4. 570, 580, 590, 600
5. 825, 830, 835

Unit 6 (p. 33)
••
1. 190, 100 + 90 + 0, one hundred ninety
2. 200 + 40 + 9, two hundred forty-nine
3. 628, six hundred twenty-eight
4. 834, 800 + 30 + 4
5. 600, 700, 800, 900
6. 490, 500, 510, 520
7. 525, 530, 535, 540
8. 660, 670, 680

Unit 6 (p. 34)
•••
1. 700 + 30 + 2, seven hundred thirty-two
2. 594, five hundred ninety-four
3. 609, 600 + 0 + 9
4. 800 + 10 + 3, eight hundred thirteen
5. 340, 345, 350
6. 800, 900, 1,000
7. 770, 780, 790
8. 460, 465, 470
9. 750, 850, 950
10. 990, 995, 1,000

Unit 6 (p. 35)
Word Problems
1. 700 + 60 + 8, seven hundred sixty-eight
2. 470, 400 + 70 + 0
3. 660, 670, 680
4. 290, 300, 310
5. 960, 965, 970
6. 720, 725, 730
7. 450, 460, 470
8. 700, 705, 710
9. 840, 845, 850
10. 350, 360, 370
11. d
12. b

Unit 7 (p. 37)
•
1. <
2. >
3. >
4. <
5. >
6. >

Unit 7 (p. 38)
••
1. >   2. >
3. <   4. <
5. <   6. >
7. <   8. <
9. =

Unit 7 (p. 39)
•••
1. <   2. >
3. <   4. =
5. <   6. <
7. >   8. >
9. <   10. <
11. >   12. =

Unit 7 (p. 40)
Word Problems
1. Sally   2. Monday
3. b   4. c

Name _____

Answer Key • Units 8–12

Unit 8 (p. 42)
●

1. 24	**2.** 32
3. 32	**4.** 50
5. 70	**6.** 84

Unit 8 (p. 43)
●●

1. 31	**2.** 33
3. 92	**4.** 55
5. 47	**6.** 48
7. 53	**8.** 60

Unit 8 (p. 44)
●●●

1. 50	**2.** 39
3. 31	**4.** 21
5. 42	**6.** 37
7. 61	**8.** 53
9. 70	**10.** 84
11. 46	**12.** 45

Unit 8 (p. 45)
Word Problems

1. 20 cards
2. 38 socks
3. 26 pennies
4. 47 cupcakes
5. c
6. d

Unit 9 (p. 47)
●

1. 65	**2.** 73
3. 59	**4.** 49
5. 90	**6.** 99

Unit 9 (p. 48)
●●

1. 72	**2.** 66
3. 98	**4.** 80
5. 81	**6.** 91
7. 103	**8.** 90

Unit 9 (p. 49)
●●●

1. 60	**2.** 71
3. 74	**4.** 81
5. 94	**6.** 100
7. 97	**8.** 80
9. 88	**10.** 80

Unit 9 (p. 50)
Word Problems

1. 31
2. 70
3. d
4. b

Unit 10 (p. 52)
●

1. 162	**2.** 235
3. 314	**4.** 150
5. 139	**6.** 202

Unit 10 (p. 53)
●●

1. 233	**2.** 378
3. 460	**4.** 301
5. 175	**6.** 46
7. 579	**8.** 605

Unit 10 (p. 54)
●●●

1. 647	**2.** 774
3. 258	**4.** 541
5. 739	**6.** 823
7. 788	**8.** 40
9. 653	**10.** 701

Unit 10 (p. 55)
Word Problems

1. 160
2. 535
3. c
4. d

Unit 11 (p. 57)
●

1. 259	**2.** 241
3. 457	**4.** 331
5. 447	

Unit 11 (p. 58)
●●

1. 284	**2.** 366
3. 546	**4.** 877
5. 898	**6.** 938
7. 915	**8.** 803

Unit 11 (p. 59)
●●●

1. 282	**2.** 384
3. 374	**4.** 519
5. 513	**6.** 660
7. 668	**8.** 530
9. 856	**10.** 533
11. 374	**12.** 417

Unit 11 (p. 60)
Word Problems

1. 871 toothpicks
2. 725 paper clips
3. b
4. c

Unit 12 (p. 62)
●

1. 39	**2.** 64
3. 7	**4.** 6
5. 71	**6.** 37

Unit 12 (p. 63)
●●

1. 8	**2.** 8
3. 6	**4.** 7
5. 9	**6.** 8
7. 5	**8.** 7

Unit 12 (p. 64)
●●●

1. 45	**2.** 22	**3.** 32
4. 71	**5.** 77	**6.** 61
7. 56	**8.** 34	**9.** 32
10. 17	**11.** 75	**12.** 46

Unit 12 (p. 65)
Word Problems

1. 81 crayons
2. 68 markers
3. a
4. c

Common Core Mathematics Grade 2 • ©2012 Newmark Learning,

Answer Key • Units 13–17

Unit 13 (p. 67)
●

1. 12   **2.** 21
3. 19   **4.** 18
5. 39   **6.** 55

Unit 13 (p. 68)
●●

1. 11   **2.** 23
3. 35   **4.** 20
5. 49   **6.** 3
7. 46   **8.** 20
9. 9   **10.** 38

Unit 13 (p. 69)
●●●

1. 21   **2.** 22
3. 10   **4.** 33
5. 31   **6.** 39
7. 26   **8.** 29
9. 22   **10.** 57
11. 36   **12.** 29

Unit 13 (p. 70)
Word Problems

1. 22 hairpins
2. 9 cupcakes
3. c
4. b

Unit 14 (p. 72)
●

1. 130   **2.** 213
3. 107   **4.** 95
5. 171   **6.** 128

Unit 14 (p. 73)
●●

1. 215   **2.** 233
3. 724   **4.** 638
5. 202   **6.** 445
7. 394   **8.** 379

Unit 14 (p. 74)
●●●

1. 223   **2.** 233
3. 285   **4.** 43
5. 413   **6.** 425
7. 365   **8.** 444
9. 605   **10.** 451
11. 664   **12.** 378

Unit 14 (p. 75)
Word Problems

1. 185 days
2. 333 pages
3. a
4. b

Unit 15 (p. 77)
●

1. 2 inches
2. 6 inches
3. 5 inches
4. 4 inches

Unit 15 (p. 78)
●●

Answers may vary.

Unit 15 (p. 79)
●●●

Answers may vary.

Unit 15 (p. 80)
Word Problems

1. Answers may vary;
approximately 7 inches
2. Answers may vary.
3. c

Unit 16 (p. 82)
●

1. 6 centimeters
2. 13 centimeters
3. 8 centimeters
4. 10 centimeters

Unit 16 (p. 83)
●●

Check students' work.
Answers may vary.

Unit 16 (p. 84)
●●●

Check students' work.
Answers may vary.

Unit 16 (p. 85)
Word Problems

Check students' work
for Problems 1 and 2.
Answers may vary.
3. b

Unit 17 (p. 87)
●

1. 20 inches
2. 15 feet
3. 14 meters

Unit 17 (p. 88)
●●

1. 2 meters
2. 30 meters
3. 13 yards
4. 18 feet
5. 3 meters

Unit 17 (p. 89)
●●●

1. 11 yards
2. 19 feet
3. 15 feet
4. 17 inches
5. 6 meters
6. 1 centimeter

Unit 17 (p. 90)
Word Problems

1. 9 inches
2. 12 centimeters
3. 9 feet
4. 2 centimeters
5. d
6. c

Answer Key • Units 18–21

Unit 18 (p. 92)
•
1. cross out 12 o'clock
2. cross out 7 o'clock
3. cross out 12:00
4. cross out 9 o'clock
5. cross out analog 4:55
6. cross out 5 o'clock

Unit 18 (p. 93)
••
1. 1:30   2. 4:30
3. 10:30   4. 1:00
5. 2:00   6. 6:45
7. 8:55   8. 12:05

Unit 18 (p. 94)
•••
1. 12:25   2. 6:10
3. 10:35   4. 3:50
5. 4:10   6. 11:50
7. 4:40   8. 7:35
9. 2:25   10. 10:05
11. 3:20   12. 12:05

Unit 18 (p. 95)
Word Problems
1. third clock
2. second clock
3. c
4. d

Unit 19 (p. 97)
•
1. 13¢
2. 33¢
3. 45¢
4. 40¢
5. $1 and 3¢

Unit 19 (p. 98)
••
1. 21¢   2. 28¢
3. 26¢   4. 47¢
5. 21¢   6. 42¢
7. 100¢   8. $2 and 11¢

Unit 19 (p. 99)
•••
1. $1.41   2. $0.65
3. $3.01   4. $2.06
5. $1.60   6. $0.41
7. $0.40   8. $1.26

Unit 19 (p. 100)
Word Problems
1. 32¢
2. 65¢
3. d
4. a

Unit 20 (p. 102)
•
1. 46
2. 5
3. 3
4. 47

Unit 20 (p. 103)
••
Check students' work.
1. 7
2. 10

Unit 20 (p. 104)
•••
Check students' work.
1. 9
2. 5
3. 9

Unit 20 (p. 105)
Word Problems
1. Check students' work.
2. 18
3. 17
4. d
5. c

Unit 21 (p. 107)
•
1. 14
2. 20
3. 4
4. 6

Unit 21 (p. 108)
••
Check students' work.
1. 15
2. 12
3. 2
4. 7

Unit 21 (p. 109)
•••
1. Check students' work.
2. 2
3. 9
4. 17
5. Check students' work.
6. 2
7. 8
8. 0

Unit 21 (p. 110)
Graph Practice
Check students' work.

Common Core Mathematics Grade 2 • ©2012 Newmark Learning, L

Answer Key • Units 22–23

Unit 22 (p. 112)
•

1. 3, 3 **2.** 4, 4
3. 5, 5 **4.** 6, 6
5. 4, 4 **6.** 8, 8

Unit 22 (p. 113)
••

1. triangle **2.** rectangle
3. pentagon **4.** hexagon
5. cube **6.** parallelogram

Unit 22 (p. 114)
•••

Check students' work.

Unit 22 (p. 115)
Word Problems

Check students' work.

Unit 23 (p. 117)
•

Check students' work.

Unit 23 (p. 118)
••

Check students' work.

Unit 23 (p. 119)
•••

Check students' work.

Unit 23 (p. 120)
Math Practice

Check students' work.

Answer Key • Fluency Practice

p. 121

1. 5 + 7 = 12, 7 + 5 = 12,
12 − 5 = 7, 12 − 7 = 5
2. 7 + 8 = 15, 8 + 7 = 15,
15 − 8 = 7, 15 − 7 = 8
3. 6 + 7 = 13, 7 + 6 = 13,
13 − 6 = 7, 13 − 7 = 6

4. 5 + 9 = 14, 9 + 5 = 14,
14 − 5 = 9, 14 − 9 = 5
5. 4 + 9 = 13, 9 + 4 = 13,
13 − 4 = 9, 13 − 9 = 4
6. 7 + 9 = 16, 9 + 7 = 16,
16 − 7 = 9, 16 − 9 = 7

7. 3 + 9 = 12, 9 + 3 = 12,
12 − 3 = 9, 12 − 9 = 3
8. 4 + 7 = 11, 7 + 4 = 11,
11 − 4 = 7, 11 − 7 = 4
9. 6 + 9 = 15, 9 + 6 = 15,
15 − 6 = 9, 15 − 9 = 6

10. 8 + 9 = 17, 9 + 8 = 17,
17 − 8 = 9, 17 − 9 = 8
11. 5 + 6 = 11, 6 + 5 = 11,
11 − 5 = 6, 11 − 6 = 5
12. 4 + 8 = 12, 8 + 4 = 12,
12 − 4 = 8, 12 − 8 = 4

p. 122

1. 32	**2.** 8
3. 28	**4.** 15
5. 8	**6.** 6
7. 18	**8.** 37
9. 47	**10.** 64
11. 74	**12.** 69
13. 23	**14.** 23

p. 123

1. 21	**2.** 22
3. 10	**4.** 33
5. 31	**6.** 39
7. 26	**8.** 29
9. 22	**10.** 57
11. 36	**12.** 29

p. 124

1. even	**2.** even
3. odd	**4.** even
5. odd	**6.** even
7. even	**8.** even
9. even	**10.** odd
11. odd	**12.** even

p. 125

1. 60	**2.** 71
3. 74	**4.** 81
5. 94	**6.** 100
7. 97	**8.** 80
9. 88	**10.** 80

p. 126

1. 99	**2.** 72
3. 90	**4.** 98
5. 100	**6.** 75
7. 94	**8.** 103
9. 98	**10.** 96

p. 127

1. <	**2.** <
3. <	**4.** >
5. =	**6.** <
7. <	**8.** <
9. <	**10.** >
11. >	**12.** >

p. 128

1. <	**2.** >
3. <	**4.** >
5. >	**6.** =
7. <	**8.** <
9. >	**10.** >
11. <	**12.** <

p. 129

1. 784	**2.** 898
3. 637	**4.** 450
5. 470	**6.** 670
7. 936	**8.** 965
9. 543	**10.** 617
11. 1,078	**12.** 871
13. 450	**14.** 646

p. 130

1. 850	**2.** 764
3. 811	**4.** 48
5. 959	**6.** 468
7. 588	**8.** 381
9. 603	**10.** 315
11. 305	**12.** 483
13. 584	**14.** 914

p. 131

1. 420	**2.** 340
3. 409	**4.** 346
5. 172	**6.** 264
7. 524	**8.** 349
9. 945	**10.** 327
11. 344	**12.** 767
13. 286	**14.** 109

Notes

Common Core Mathematics Grade 2 • ©2012 Newmark Learning, Ll